# Polarization Dynamics of Mode-Locked Fiber Lasers

This book provides a comprehensive review of the latest research on the science, technology, and applications of mode-locked fiber lasers generating pulse trains with the evolving state of polarization at time scales ranging from a few pulse widths to 10,000 laser cavity round-trip times. It supports readers with a timely source of information on the current novel scientific concepts, and cost-effective schematics, in addition to an overview of the feasible applications.

The book aims to demonstrate, for the nonlinear science community, a newly emerging field of nonlinear science, and so stimulates the development of new theoretical approaches and opens new horizons for the photonics community by pushing the boundaries of the existing laser systems toward new applications. The new classes of optical sources and photonic devices explored in this book will have relevant applications in other fields, including medicine, bio-photonics, metrology, and environmental safety.

Features

- Provides a cutting-edge review of the latest emerging science, technology and applications in the field.
- Tackles a topic with fast-growing interest in the USA, Europe, and China.
- Explores the simple and cheap design and tests of lasers, and outlines the feasible applications.

**Sergey V. Sergeyev** received his MSc and PhD degrees in optics and laser physics from Belarusian State University (BSU), Minsk, in 1985 and 1991, respectively. He has 34 years of academic experience and three years of industrial experience in the field of polarization spectroscopy, telecom, laser physics, and nonlinear fiber optics. He is now an Associate Professor at the Aston Institute of Photonic, Technologies (AIPT) at Aston University, UK. He has published over 200 journal and conference papers, four book chapters, and filed seven patent applications. He has recently supervised of six EU and UK projects covering the polarization phenomena in spectroscopy, fiber optics, telecom, and laser physics.

**Chengbo Mou** received his B.Eng. degree in electronic science and technologies (optoelectronics) from Tianjin University, China, in 2004, in addition to his M.Sc. degree in photonics and optoelectronic devices from the University of St Andrews, U.K., in 2005, and a Ph.D. degree in photonics from Aston University in 2012. He then worked as an Industrial Research Fellow with Aston University. In 2016, he joined in the Key Laboratory of Specialty Fiber Optics and Optical Access Networks at Shanghai University as a Full Professor. His research interests include nanophotonics, nanomaterial-based nonlinear photonic devices, ultrafast fiber lasers, novel type of mode locked lasers, and nonlinear applications of advanced fiber grating devices. He is the recipient of National High Level Oversea Recruitment programme of China, the Young Eastern Scholar Fellowship from the Shanghai Institute of Higher Learning. He has published over 100 journal and conference papers, three book chapters. He has been extensively involved in projects supported by Natural Science Foundation of China, Natural Science Foundation of Shanghai, Ministry of Education China, and MOST China.

# Polarization Dynamics of Mode-Locked Fiber Lasers

## Science, Technology, and Applications

Edited by

Sergey V. Sergeyev
Chengbo Mou

CRC Press
Taylor & Francis Group
Boca Raton  London  New York

CRC Press is an imprint of the
Taylor & Francis Group, an **informa** business

First edition published 2023
by CRC Press
6000 Broken Sound Parkway NW, Suite 300, Boca Raton, FL 33487-2742

and by CRC Press
4 Park Square, Milton Park, Abingdon, Oxon, OX14 4RN

*CRC Press is an imprint of Taylor & Francis Group, LLC*

*Library of Congress Cataloging-in-Publication Data*
Names: Sergeyev, Sergey V., editor. | Mou, Chengbo, editor.
Title: Polarization dynamics of mode-locked fiber lasers : science,
technology, and applications / edited by Sergey V. Sergeyev and Chengbo Mou.
Description: Boca Raton : CRC Press, 2023. | Includes bibliographical references and index. |
Summary: "This book provides a comprehensive review of the latest research related to the science, technology and applications of mode-locked fiber lasers. It provides readers with a timely source of information on the recent novel scientific concepts, cost-effective schematics, in addition to an overview of the feasible applications such as in LIDAR in autonomous cars and plant vegetation and environment monitoring. It also contains a review of recent scientific findings in the field of VOFCs, which demonstrate for the nonlinear science community a newly emerging field of nonlinear science, and so stimulates the development of new theoretical approaches and opens new horizons for the photonics community by pushing boundaries of the existing laser systems in terms of optical bandwidth and acquisition speed and resulting new applications. This book is a valuable guide for researchers in photonics and fundamental and nonlinear science, in addition to the fields of sensing and chemical monitoring. The new classes of optical sources and photonic devices explored in this book will be of relevance with applications to other fields including medicine, bio-photonics, metrology, and environmental safety"-- Provided by publisher.
Identifiers: LCCN 2022048150 | ISBN 9781032064505 (hardback) | ISBN 9781032074061 (paperback) | ISBN 9781003206767 (ebook)
Subjects: LCSH: Raman lasers. | Mode-locked lasers. | Polarization (Light) | Lasers--Industrial applications.
Classification: LCC TA1699 .P65 2023 | DDC 621.36/6--dc23/eng/20230120
LC record available at https://lccn.loc.gov/2022048150

ISBN: 978-1-032-06450-5 (hbk)
ISBN: 978-1-032-07406-1 (pbk)
ISBN: 978-1-003-20676-7 (ebk)

DOI: 10.1201/9781003206767

Typeset in Times
by SPi Technologies India Pvt Ltd (Straive)

# Contents

List of Contributors ................................................................................................. vii

**Chapter 1**   Polarization Dynamics in Mode-Locked Fiber Lasers ........................ 1

*Sergey V. Sergeyev, Chengbo Mou, Hani J. Kbashi and Stanislav A. Kolpakov*

**Chapter 2**   Recent Development of Polarizing Fiber Grating Based Mode-Locked Fiber Laser ................................................................... 69

*Zinan Huang, Yuze Dai, Qianqian Huang, Zhikun Xing, Lilong Dai, Weixi Li, Zhijun Yan and Chengbo Mou*

**Chapter 3**   Polarization and Color Domains in Fiber Lasers ............................ 103

*Yichang Meng, Ahmed Nady, Georges Semaan, A. Komarov, M. Kemel, M. Salhi and F. Sanchez*

**Chapter 4**   Dual-Output Vector Soliton Fiber Lasers ....................................... 123

*Michelle Y. Sander and Shutao Xu*

**Chapter 5**   Vector Solitons Formed in Linearly Birefringent Single Mode Fibers ............................................................................................... 143

*X. Hu, J. Guo and D. Y. Tang*

**Chapter 6**   Vector Solitons in Figure-Eight Fiber Lasers .................................. 169

*Min Luo and Zhi-Chao Luo*

**Chapter 7**   Polarization Dynamics of Mode-Locked Fiber Lasers with Dispersion Management .................................................................. 189

*Tao Zhu and Lei Gao*

**Chapter 8**   Dual-Wavelength Fiber Laser for 5G and Lidar Applications ......... 205

*Hani J. Kbashi and Vishal Sharma*

**Index** ................................................................................................................. 223

Contents

# List of Contributors

**Lilong Dai**
Key Lab of Specialty Fiber Optics
and Optical Access Network,
Shanghai Institute for Advanced
Communication and Data Science,
Joint International Research
Laboratory of Specialty Fiber Optics
and Advanced Communication,
Shanghai University
Shanghai, China

**Yuze Dai**
National Engineering Research
Center for Next Generation Internet
Access System, School of Optical
and Electronic Information,
Huazhong University of Science and
Technology
Wuhan, China

**Lei Gao**
Key Laboratory of Optoelectronic
Technology & Systems (Ministry of
Education), Chongqing University
Chongqing, China

**J. Guo**
School of Physics and Electronic
Engineering, Jiangsu Normal
University
Xuzhou, China

**X. Hu**
School of Electrical and Electronic
Engineering, Nanyang Technological
University
Singapore

**Qianqian Huang**
Key Lab of Specialty Fiber Optics
and Optical Access Network,
Shanghai Institute for Advanced

Communication and Data Science,
Joint International Research
Laboratory of Specialty Fiber Optics
and Advanced Communication,
Shanghai University
Shanghai, China

**Zinan Huang**
Key Lab of Specialty Fiber Optics
and Optical Access Network,
Shanghai Institute for Advanced
Communication and Data Science,
Joint International Research
Laboratory of Specialty Fiber Optics
and Advanced Communication,
Shanghai University
Shanghai, China

**Hani J. Kbashi**
Aston Institute of Photonic
Technologies, College of Engineering
and Physical Sciences, Aston
University
Birmingham, United Kingdom

**M. Kemel**
Laboratoire de Photonique d'Angers
E.A. 4464, Université d'Angers
Angers, France

**Stanislav A. Kolpakov**
Aston Institute of Photonic
Technologies, College of Engineering
and Physical Sciences, Aston
University
Birmingham, United Kingdom

**A. Komarov**
Laboratoire de Photonique d'Angers
E.A. 4464, Université d'Angers
Angers, France

Institute of Automation and
Electrometry, Russian Academy of
Sciences, Acad. Koptyug Pr. 1
Novosibirsk, Russia

**Weixi Li**
Key Lab of Specialty Fiber Optics
and Optical Access Network,
Shanghai Institute for Advanced
Communication and Data Science,
Joint International Research
Laboratory of Specialty Fiber Optics
and Advanced Communication,
Shanghai University
Shanghai, China

**Min Luo**
Guangdong Provincial Key Laboratory
of Nanophotonic Functional
Materials and Devices & Guangzhou
Key Laboratory for Special Fiber
Photonic Devices and Applications,
South China Normal University
Guangzhou, Guangdong, China

**Zhi-Chao Luo**
Guangdong Provincial Key Laboratory
of Nanophotonic Functional
Materials and Devices & Guangzhou
Key Laboratory for Special Fiber
Photonic Devices and Applications,
South China Normal University
Guangzhou, Guangdong, China

**Yichang Meng**
Laboratoire de Photonique d'Angers
E.A. 4464, Université d'Angers
Angers, France
Hebei University of Science and
Technology
Shijiazhuang, China

**Chengbo Mou**
Key Lab of Specialty Fiber Optics
and Optical Access Network,
Shanghai Institute for Advanced

Communication and Data Science,
Joint International Research
Laboratory of Specialty Fiber Optics
and Advanced Communication,
Shanghai University
Shanghai, China

**Ahmed Nady**
Laboratoire de Photonique d'Angers
E.A. 4464, Université d'Angers
Angers, France
Beni-Suef University, Faculty of
Science
Beni Suef, Egypt

**M. Salhi**
Laboratoire de Photonique d'Angers
E.A. 4464, Université d'Angers
Angers, France

**F. Sanchez**
Laboratoire de Photonique d'Angers
E.A. 4464, Université d'Angers
Angers, France

**Michelle Y. Sander**
Department of Electrical and Computer
Engineering and BU Photonics
Center, Boston University
Boston, MA, USA
Division of Materials Science and
Engineering, Boston University
Brookline, MA, USA

**Georges Semaan**
Laboratoire de Photonique d'Angers
E.A. 4464, Université d'Angers
Angers, France

**Sergey V. Sergeyev**
Aston Institute of Photonic
Technologies, College of Engineering
and Physical Sciences, Aston
University
Birmingham, United Kingdom

**Vishal Sharma**
School of Engineering and Applied
    Science, Aston University
Birmingham, UK
SBS State Technical Campus, IKG
    Punjab Technical University
Punjab, India

**D. Y. Tang**
School of Electrical and Electronic
    Engineering, Nanyang Technological
    University
Singapore

**Zhikun Xing**
National Engineering Research Center
    for Next Generation Internet Access
    System, School of Optical and
    Electronic Information, Huazhong
    University of Science and Technology
Wuhan, China

**Shutao Xu**
Department of Electrical and Computer
    Engineering and BU Photonics
    Center, Boston University
Boston, MA, USA

**Zhijun Yan**
National Engineering Research
    Center for Next Generation Internet
    Access System, School of Optical
    and Electronic Information,
    Huazhong University of Science and
    Technology
Wuhan, China

**Tao Zhu**
Key Laboratory of Optoelectronic
    Technology & Systems (Ministry of
    Education), Chongqing University
Chongqing, China

# 1 Polarization Dynamics in Mode-Locked Fiber Lasers

*Sergey V. Sergeyev, Chengbo Mou, Hani J. Kbashi and Stanislav A. Kolpakov*

## CONTENTS

1.1  Introduction ..................................................................................................... 1
1.2  Fundamental Soliton Polarization Dynamics (Experiment) ........................... 2
    1.2.1  Experimental Set-Up .......................................................................... 2
    1.2.2  Experimentally Observed Fundamental Soliton's Polarization
        Attractors .............................................................................................. 4
1.3  Vector Multipulsing Soliton Dynamics (Experiment) .................................... 5
1.4  Polarization Dynamics of Bound State Solitons (Experiment) ....................... 9
1.5  Vector Soliton Rain (Experiment) ................................................................ 18
1.6  Vector Bright-Dark Rogue Waves (Experiment) .......................................... 21
1.7  Vector Resonance Multimode Instability (Experiment) ............................... 23
1.8  Vector Harmonic Mode-Locking (Experiments) ......................................... 26
1.9  Vector Model of an Erbium-Doped Fiber Laser ........................................... 30
    1.9.1  Semiclassical Equations .................................................................... 30
    1.9.2  Reducing the Complexity of the Semiclassical Model .................... 32
1.10  Spiral Polarization Attractor (Theory) ........................................................ 37
1.11  Interplay Between Polarization Hole Burning and In-Cavity
    Birefringence (Theory) .................................................................................. 41
1.12  Vector Soliton Rain (Theory) ...................................................................... 44
1.13  Vector Bright-Dark Rogue Waves (Theory) ................................................ 47
1.14  Vector Resonance Multimode Instability (Theory) ..................................... 49
1.15  Vector Harmonic Mode-Locking (Theory) ................................................. 53
1.16  Self-Pulsing in Fiber Lasers (Theory) ......................................................... 57
Acknowledgement .................................................................................................. 63
References ............................................................................................................... 63

## 1.1  INTRODUCTION

Dissipative vector solitons (DVSs) in mode-locked fiber lasers compose a train of stabilized ultrashort pulses with the specific shape and state of polarization (SOP) driven by a complex interplay between the effects of gain/loss, dispersion, nonlinearity, and linear and circular birefringence (Cundiff et al., 1999; Haus et al., 1999;

DOI: 10.1201/9781003206767-1

Tang et al., 2008; Zhao et al., 2008; Zhang et al., 2009; Mou et al., 2011; Grelu and Akhmediev, 2012; Boscolo et al., 2014; Sergeyev, 2014; Sergeyev et al., 2014). Given the SOP of the solitons can be locked or evolved at different time scales, the DVSs' stability is an important issue to be addressed in the context of applications in metrology (Udem et al., 2002; Zhao et al., 2018; Pupeza et al., 2021), spectroscopy (Mandon et al., 2009; Picqué and Hänsch, 2019) and high-speed fiber-optic communication (Hillerkuss et al., 2011; Geng et al., 2022). In addition, the flexibility in the control of dynamic SOPs is of interest for trapping and manipulation of atoms and nanoparticles (Spanner et al., 2001; Jiang et al., 2010; Tong et al., 2010; Misawa, 2016; MacPhail-Bartley et al., 2020), and control of magnetization (Kanda et al., 2011; Kimel and Li, 2019).

The stability and evolution of vector solitons at a time interval from just a few to thousands of cavity round trips is defined by asymptotic states (attractors) which the laser SOP approaches at a long time scale, *viz.* fixed point, periodic, quasi-periodic, and chaotic dynamics. High signal-to-noise ratio (>40 dB) measurement in the case of mode-locked lasers and application of a polarimeter, gives an opportunity for the direct observation of attractors embedded in 3D space in terms of the Stokes parameters $S_1$, $S_2$, and $S_3$ (Mou et al., 2011; Sergeyev, 2014; Sergeyev et al., 2014). In this section, we review our recent experimental study of the single soliton polarization dynamics in Er-doped mode-locked fiber lasers at time scales from 1 to 100,000 roundtrips. To characterize the dynamics theoretically, we review our new vector model in Sections 1.9–1.11.

## 1.2  FUNDAMENTAL SOLITON POLARIZATION DYNAMICS (EXPERIMENT)

### 1.2.1  EXPERIMENTAL SET-UP

Figure 1.1a illustrates the Er-doped fiber laser (EDFL) mode-locked by the carbon nanotubes (CNT) saturable absorber (SA). The EDFL gain medium has of 2 m of high-concentrated Er-doped fiber (EDF Er80-8/125 from Liekki) which is pumped by 980 nm laser diode (LD) through 980 nm/1,550 nm wavelength division multiplexing (WDM). The external and in-cavity polarization controllers are used to adjust the pump wave and lasing states of polarization. An optical isolator (OISO) provides a unidirectional lasing signal propagation in the laser cavity. The output coupler (OUTPUT C) redirects of 10% of light outside the cavity. The total length of the laser cavity is 7.83 m with an average anomalous dispersion (group velocity dispersion (GVD) parameter for erbium fiber $\beta_{2,\text{EDF}} = -19.26$ fs$^2$/mm) that will result in soliton output. To characterize the output lasing, auto-correlator (Pulsecheck), oscilloscope (Tektronix), optical spectrum analyzer (ANDO AQ6317B), and in-line polarimeter (Thorlabs, IPM5300) are used. The EDFL is pumped at 178 mW, resulting in 3 mW of averaged lasing power.

Figure 1.1b shows an output optical spectrum of the output lasing signal centered at 1,560 nm and having a full-width half-maximum (FWHM) spectral bandwidth of 3.72 nm. The Kelly sidebands indicate the fundamental soliton shape of the output pulses. A pulse train has a period of 38.9 ns or the repetition rate of 25.7 MHz (Figure 1.1c). The measured autocorrelation trace corresponding to the pulse duration of 583 fs is shown in Figure 1.1d.

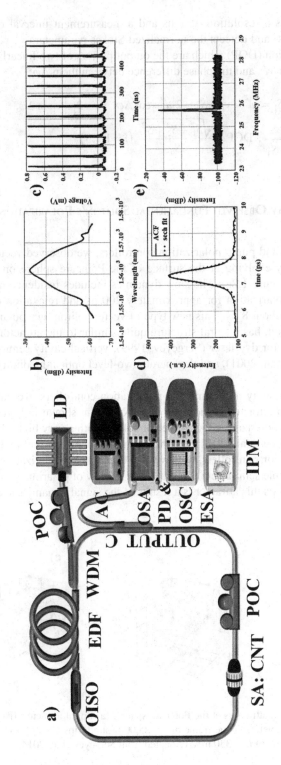

**FIGURE 1.1** Experimental set-up. EDF, high concentration erbium–doped fiber; SM, single-mode fiber with anomalous dispersion; POC, polarization controller; WDM, a wavelength division multiplexing coupler; OISO, an optical isolator; CNT, a fast saturable absorber (carbon nanotubes); OUTPUT C, an output coupler; AC, autocorrelator; OSA, optical spectrum analyzer; PD, photodetector; OSC, oscilloscope; ESA, electrical spectrum analyzer; IPM, inline polarimeter. (Adapted from Sergeyev et al. 2014).

The polarimeter has a resolution of 1 μs and a measurement interval of 1 ms (25–25,000 round trips) and detects the normalized Stokes parameters $s_1$, $s_2$, $s_3$ and the degree of polarization (DOP) which are the output powers of two linearly cross-polarized SOPs, $|u|^2$ and $|v|^2$, and the phase difference between them $\Delta\varphi$:

$$S_0 = |u|^2 + |v|^2, \ S_1 = |u|^2 - |v|^2, \ S_2 = 2|u||v|\cos\Delta\varphi, \ S_3 = 2|u||v|\sin\Delta\varphi,$$

$$S_i = \frac{S_i}{\sqrt{S_1^2 + S_2^2 + S_3^2}}, \ \mathrm{DOP} = \frac{\sqrt{S_1^2 + S_2^2 + S_3^2}}{S_o}, (i = 1,2,3) \tag{1.1}$$

### 1.2.2 EXPERIMENTALLY OBSERVED FUNDAMENTAL SOLITON'S POLARIZATION ATTRACTORS

By adjusting in-cavity and pump polarization controllers, we observed vector solitons with SOPs slowly evolving at the surface of the Poincaré sphere on a double spiral trajectory (Figure 1.2a). The slow dynamics includes residence near the orthogonal states of polarization for approximately 200 μs and relaxation oscillations with a period of about 8 μs. This new type of vector soliton, *viz.* polarization precessing vector soliton, has a spiral structure quite similar to the attractors demonstrated theoretically for dye laser (Sergeyev, 1999), vertical-cavity semiconductor laser (Willemsen et al., 2001), and degenerate two-level optical medium (Byrne et al., 2003).

By tuning the intra-cavity and pump LD polarization controllers, we have also observed a polarization attractor in the form of a fixed point shown in Figure 1.2b that corresponds to the polarization locked vector soliton with a very high degree of polarization of 92% (Mou et al., 2011). By increasing the pump current to 330 mA and tuning the polarization controllers, we have also found a new polarization attractor in the form of a double semi-circle (Figure 1.2c). In view of polarimeter's photo-detector resolution of 1 μs this attractor is a results of the signal dynamics averaging over 25 round trips.

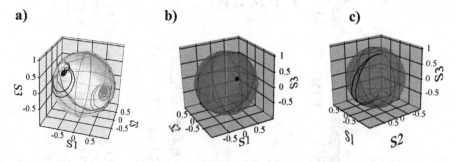

**FIGURE 1.2** Polarization attractors at the Poincaré sphere: (a) Spiral attractor; (b) Locked SOP; (c) Double arc. Parameters: Time frame of 25–25,000 round trips (1 μs–1 ms); (a), (b) Pump current $I_p = 310$ mA; (c) $I_p = 330$ mA. (Adapted from Sergeyev et al. 2014.)

## 1.3 VECTOR MULTIPULSING SOLITON DYNAMICS (EXPERIMENT)

By tuning an in-cavity polarization controller (POC) and POC for the pump laser (Figure 1.1a), we have found a new type of vector solitons with precessing SOPs for multipulsing operations at a time scale of 25–25,000 roundtrips (Sergeyev et al., 2012). In addition to a slow polarimeter IPM5300 with 1 μs resolution we used a fast polarimeter inline polarimeter (OFS TruePhase® IPLM)[13] optimized for the high-speed operation (Figure 1.3) (Tsatourian et al. 2013a). The polarimeter comprises four tilted fiber gratings (TFBGs) inscribed in the core of polarization-maintaining (PM) fiber. Each TFBG scatters 1% of incoming light on the detector and four detectors' voltages were recorded simultaneously by oscilloscope (Tektronix DPO7254). The polarimeter has 3 dB bandwidth of 550 MHz with a maximum DOP error of around ±4%. The self-calibration procedure has been used to convert the detectors signals voltage to Stokes parameters.

In the experiment, the pump current has been changed from 306 mA to 355 mA, and the in-cavity polarization and pump polarization controller have been adjusted. All autocorrelation traces have been averaged over 16 samples to mitigate the autocorrelator sensitivity to the input SOP.

With the pump current of 306 mA, double-pulsing was observed (Figure 1.4a–f). The output optical spectrum shown in Figure 1.4a is centered at 1,560 nm with Kelly sidebands indicating the fundamental soliton shape. The pulses doubling shown in Figure 1.4b is the result of the interplay between the laser cavities' bandwidth constraints and the energy quantization associated with the resulting mode-locked pulses (Li et al., 2010). With increased pump power, the peak power increases, and the pulse width (inversely proportional to the spectral bandwidth) decreases according to the area theorem (Li et al., 2010). The increase in the mode-locked spectral bandwidth is limited by the gain bandwidth of the cavity. This constraint can be overcome by a further single pulse split into two pulses with energy divided between two pulses within the gain bandwidth. Given this, a double pulsing with the period $T = 38.9$ ns, pulse width $T_p = 247$ fs, and output power $I \approx 0.55$ mW was observed (Figure 1.4b). Averaging over 16 autocorrelation traces is enough to obtain a smooth soliton autocorrelation trace (Figure 1.4c). The slow dynamics recorded by IPM5300 polarimeter demonstrates that the anti-phase

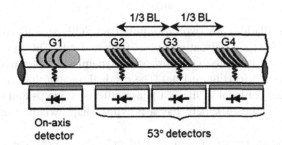

**FIGURE 1.3** Inline polarimeter: OFS TruePhase® IPLM. (Adapted from Tsatourian et al. 2013a.)

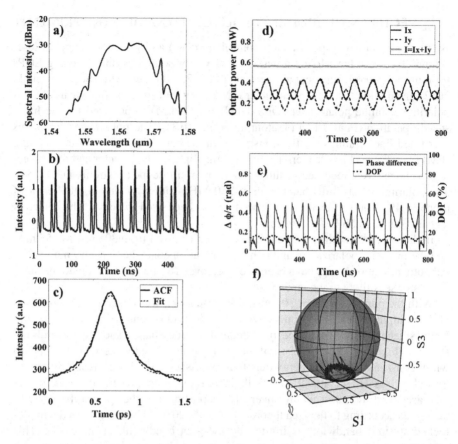

**FIGURE 1.4** Vector soliton with a slowly evolving state of polarization for two-pulse operation. (a) Output optical spectrum; (b) Single pulse train; (c) Measured autocorrelation trace. Polarization dynamics in the time frame of 25–25,000 round trips (1 μs–1 ms) in terms of (d) Optical power of orthogonally polarized modes $I_x$ (solid line) and $I_y$ (dashed line), total power $I = I_x + I_y$ (dotted line); (e) Phase difference and degree of polarization; and (f) Stokes parameters at Poincaré sphere. Parameters: Pump current $I_p$ = 306 mA, Period $T$ = 38.9 ns, Pulse width $T_p$ = 247 fs, Output power $I \approx 0.55$ mW. (Reprinted with permission from Sergeyev et al., 2012.)

dynamics of oscillations for two cross-polarized SOPs lead to continuous wave (cw) operation for the total output power (Figure 1.4d). The low value of DOP oscillations at 12% indicate that polarization dynamics is faster than the polarimeter's resolution of 1 μs (Figure 1.4e). The residual trace of the fast oscillations in Figure 1.4e takes the form of the fast phase difference jumps and so polarization attractor at the Poincaré sphere has a polyline shape winding around a circle (Figure 1.4f).

When the pump power current of 320 mA, the optical spectrum shown in Figure 1.5a preserves the soliton shape and five-pulse soliton dynamics with period $T = 38.9$ ns, pulse width $T_p = 292$ fs, output power $I \approx 0.65$ mW emerges (Figure 1.5b).

As follows from Figure 1.5c, the fast dynamics of the output SOP results in nons-moothed autocorrelation trace even after averaging over 16 samples. Given the powers of two cross-polarized SOPs slightly deviate from the anti-phase dynamics, the total output power is oscillating with a small amplitude (Figure 1.5d). Similar to the previous case (Figure 1.4e), DOP is oscillating around the low value of 30% that also indicates the presence of fast SOP oscillations faster than 1 μs that can be also justified by phase difference dynamics (Figure 1.5e). Combination of the fast dynamics in the form of the phase jumps between cross polarized SOPs and slow SOP

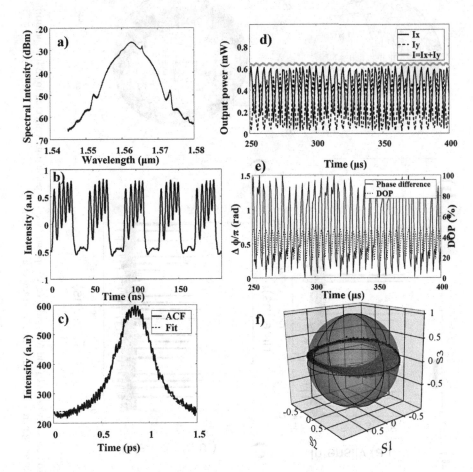

**FIGURE 1.5** Vector soliton with slowly evolving state of polarization for five-pulse operation. (a) Output optical spectrum; (b) Single pulse train; (c) Measured auto-correlation trace. Polarization dynamics in the time frame of 25–25,000 round trips (1 μs–1 ms) in terms of (d) Optical power of orthogonally polarized modes $I_x$ (solid line) and $I_y$ (dashed line), total power $I = I_x + I_y$ (dotted line); (e) Phase difference and degree of polarization; and (f) Stokes parameters at Poincaré sphere. Parameters: Pump current $I_p = 320$ mA, Period $T = 38.9$ ns, Pulse width $T_p = 292$ fs, Output power $I \approx 0.65$ mW. (Reprinted with permission from Sergeyev et al. 2012.)

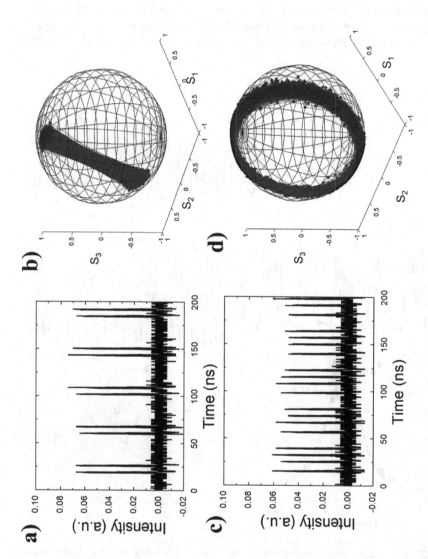

**FIGURE 1.6** Vector soliton for two- (a, b) and four-pulse (c, d) operation. (a, c) pulse train collected from four polarimeter photodetectors, (b, d) polarization attractors at Poincaré sphere. (Adapted from Tsatourian et al., 2013b.)

evolution takes the form of a polyline with an outline in the form of a circle at the Poincaré sphere (Figure 1.5f).

In this experiment with the fast polarimeter, OFS TruePhase® IPLM, the pump current was 355 mA, and the in-cavity polarization and pump polarization controllers have been tuned to obtain the polarization attractors shown in Figure 1.6. The two-pulses polarization dynamics shown in Figure 1.6a, b takes the form of polarization switching between cross-polarized SOPs with the period equal to the pulse roundtrip in the laser cavity. A four-pulse soliton operation is shown in Figure 1.6c, d. The polarization dynamics shown in Figure 1.6d demonstrates slow pulse-to-pulse evolution of the laser SOP with the period of 335 ns corresponding to eight round trips of the laser cavity. Unlike the previous case, the vector soliton shown in Figure 1.6d demonstrates slow cyclic SOP evolution with the circle trajectories at the Poincaré sphere (Figure 1.6d). It has been also found for two cases that DOP is oscillating around 90% that indicates the SOPs of the two adjacent pulses within one round trip are the same (Tsatourian et al., 2013b).

## 1.4 POLARIZATION DYNAMICS OF BOUND STATE SOLITONS (EXPERIMENT)

Unlike multipulsing with the pulse separation of nanoseconds, the bound states originate from short-range interaction through the overlapping of solitons tails or soliton-dispersive wave interaction results and results in double pulses with the spacing of the few pulse widths and phase differences of $0, \pi$ or $\pm \pi /2$. The tightly BS solitons have been experimentally observed in fiber lasers with different mode-locking techniques, including nonlinear polarization rotation (Wang et al., 2020), figure-of-eight (Seong and Kim, 2002), carbon nanotubes (CNT) (Wu et al., 2011; Gui et al., 2013; Mou et al., 2013; Tsatourian et al., 2013a) and graphene-based mode-locked fiber lasers (Li et al., 2012). In addition, various types of different bound states have been studied theoretically and experimentally, including vibrating bound states, oscillating bound states (Soto-Crespo et al., 2007) bound states with flipping and independently evolving phase (Zavyalov et al., 2009; Ortac et al., 2010). Stable bound states – soliton molecules can be used for coding and transmission of information in high-level modulation formats when multiple bits are transmitted per clock period, increasing capacity of communication channels beyond binary coding limits (Rohrmann et al., 2012). In this section, we review our recent experimental results on a new type BS solitons, namely vector BS soliton with evolving states of polarization (Mou et al., 2013; Tsatourian et al., 2013a). All results are obtained based on slow (IPM5300) and fast inline polarimeters.

The BS solitons with different pulse separation and a phase shift can be found from optical spectra analysis. For two-soliton BS with pulse separation $\tau$ and a phase shift $\varphi$, the amplitude takes the form $f(t) + f(t + \tau) \exp(i\varphi)$ and so the optical spectral power can be found as follows (Tsatourian et al., 2013a):

$$S(v) = \left| F(v) + F(v) \exp(-i2\pi[\tau v + \varphi]) \right|^2 = 2 \left| F(v) \right|^2 (1 + \cos(2\pi v\tau - \varphi)), \quad (1.2)$$

Here $F(v) = FFT(f(t))$ is the Fourier transform. The results are found in Figure 1.7a–d. As follows from Figure 1.7a–d, the optical spectrum is modulated with the frequency $\Delta v = 1/\tau$, symmetry of spectrum depends on the phase shift, and the minimum of the spectral power is zero. Though some authors associate the loss of spectral fringes' contrast with so-called vibrating solitons (Soto-Crespo et al., 2007), the interleaving of two two-pulse bound states with the phase shifts (0 and $-\pi/2$ or $\pi$ and $\pi/2$) supported by harmonic mode locking can lead to reduced fringes' contrast (Figure 1.7e, f (Tsatourian et al., 2013a)). For anomalous dispersion, pulse shape is a hyperbolic-secant-squared with the time-bandwidth product of 0.315. So, the pulse width $\Delta T$ and separation $\tau$ can be found from an optical spectrum as follows

$$\Delta T = \frac{0.315 n \lambda^2}{c \Delta \lambda}, \tau = \Delta T/N. \tag{1.3}$$

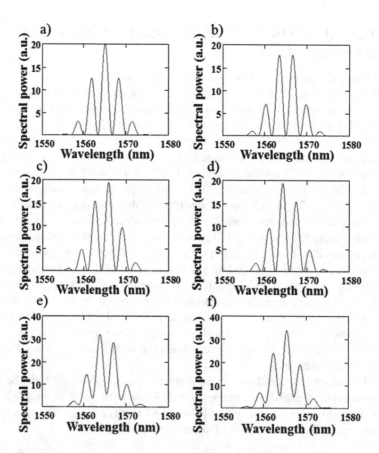

**FIGURE 1.7**    Spectra of two-soliton bound states with phase shift of: (a) 0; (b) $\pi$; (c) $-\pi/2$; (d) $\pi/2$; (e, f) interleaving of two-soliton bound states with phase shifts of: (e) 0 and $-\pi/2$, (f) $\pi$ and $\pi/2$. (Reprinted with permission from Tsatourian et al., 2013a.)

Here $n = 1.44$ is a refractive index of silica fiber, $\lambda$ is the central wavelength in optical spectrum, $c$ is the speed of light, and $N$ is the number of minima in optical spectrum.

During the experiment with slow IPM5300 polarimeter, the pump current has been varied from 240 mA to 355 mA while both intra-cavity polarization and pump polarization controllers have been adjusted to reveal the polarization attractors shown in Figures 1.8–1.11. To reduce the sensitivity of the autocorrelator to the input SOP, all autocorrelation traces were averaged over 16 samples.

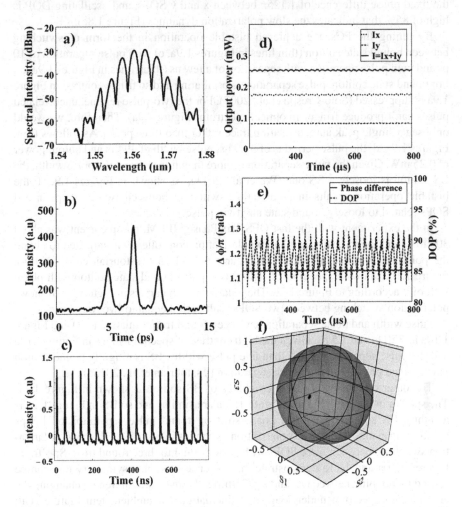

**FIGURE 1.8**  Polarization locked vector bound state soliton. (a) Output optical spectrum; (b) Measured autocorrelation trace; (c) Single pulse train. Polarization dynamics in the time frame of 25–25,000 round trips (1 µs–1 ms) in terms of (d) Optical power of orthogonally polarized modes $I_x$ (dotted line) and $I_y$ (dashed line), total power $I = I_x + I_y$ (solid line); (e) Phase difference and degree of polarization; and (f) Stokes parameters at Poincaré sphere. Parameters: Pump current $I_p = 240$ mA, Period $T = 38.9$ ns, Pulse width $T_p = 494$ fs, Output power $I = 0.25$ mW, Phase difference $\Delta\varphi \approx 1.125\pi$. (Reprinted with permission from Mou et al., 2013.)

Figure 1.8 shows the experimental results for pump current of 240 mA. The output optical spectrum with fringes shown in Figure 1.8a justifies the formation of a bound state soliton with $\pi$ phase difference (Mou et al., 2013). The three-peaks autocorrelation structure shown in Figure 1.8b has a peaks' separation of 2.5 ps, that is five times of the pulse duration of 494 fs and so presents the evidence of the tightly bound soliton (Wu et al., 2011). The pulse train with the repetition rate of 25.7 MHz demonstrates that the laser is operating in a fundamental soliton regime (Figure 1.8c). The output power is oscillating with a small amplitude (Figure 1.8d), the output SOP has the fixed phase difference of $1.125\pi$ between x and y SOPs, and oscillating DOP is high of 85% that indicates the slow polarization dynamics (Figure 1.8e).

By tuning the PCs, we achieved bistable operation in the form of switching between bound state soliton (thin lines in Figures 1.9a, c) with pulse separation of 10 ps and twin pulse operation with separation of a few ns (thick lines in Figures 1.9a, c). For bound state soliton, pulse separation is oscillating and so fringe contrast in Figure 1.9a is suppressed (Soto-Crespo et al., 2007). For the two-pulse regime, uncorrelated pulses can't produce fringes in optical spectrum (Figure 1.9a). Therefore, we would only see a single peak autocorrelation trace rather than three peaks. As follows from Figures 1.9c, d, the pulse period is of 38.9 ns, pulse width of 494 fs and output power of 0.25 mW. Given the pulse separation is more than five times of the pulse width, the BS soliton is loosely a BS one (Wu et al., 2011). As shown in Figures 1.9e, f, the bistable operation results in polarization switching between two cross polarized SOPs related to loosely bound state and twin pulse.

In the experiment with the fast OFS TruePhase® IPLM, pump current was about 300 mA, and the in-cavity and pump polarization controllers was adjusted to obtain the polarization attractors shown in Figures 1.10–1.13 (Tsatourian et al., 2013a). Figure 1.10a shows a spectrum of tightly two-pulse bound state soliton with phase shift of $\pi$ according to Figure 1.7b. The polarization dynamics in Figure 1.10b shows polarization switching between two SOPs with period equal to two round trips.

Pulse width and pulse separation have been found from Equation (1.3) and Figure 1.10a as 370 fs and 1.5 ps. Given the high contrast of spectral fringes in Figure 1.10a and the pulse separation is less than five pulse widths, BS is a tightly bound soliton having fixed phase shift and pulse separation (Tsatourian et al., 2013a).

The other type of polarization dynamics of BS soliton is shown in Figure 1.11. The spectra in Figure 1.11a and Figure 1.10a are similar and so BS in Figure 1.11 is a tightly BS soliton with fixed phase shift of $\pi$ and pulse separation of 1.5 ps (Tsatourian et al., 2013a). The polarization dynamics in Figure 1.11b shows polarization switching between three SOPs with period equal to three round trips. Spectra in Figures 1.10a and 1.11a demonstrate the presence of slight asymmetry that can be caused by hopping between $\pi$- and $-\pi/2$-shifted bound states driven by changing the erbium gain spectrum under long-term fluctuations of ambient temperature (Gui et al., 2013). High contrast of fringes and small asymmetry of spectrum justifies that lifetime in $\pi$-shifted BS is much longer than lifetime in $-\pi/2$-shifted BS.

Finally, the spectrum in Figure 1.12a demonstrates close to the $-\pi/2$-shifted tightly BS with pulse width of 370 fs and pulse separation of 1.5 ps. The SOP evolution comprises a combination of switching between three SOPs with a precession on a circle located on Poincaré sphere with the periods of 3 and 20 round trips (Figure 1.12b).

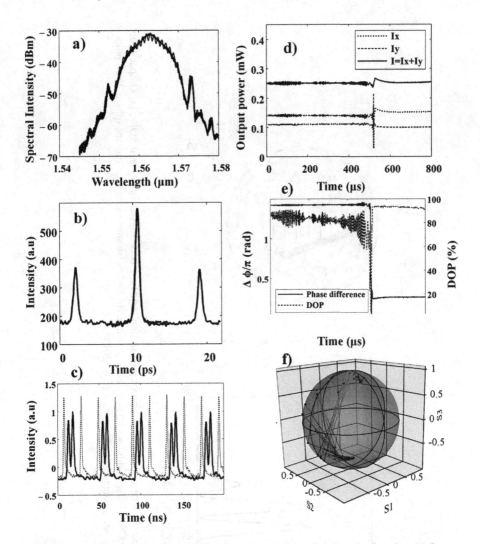

**FIGURE 1.9**  Vector soliton switching between loosely bound state and twin pulse. (a) Output optical spectrum of bound soliton (thin solid line) and twin pulse operation (thick solid line); (b) Measured autocorrelation trace for bound soliton; (c) Single pulse train of bound soliton (dashed line) and twin pulse operation (solid line). Polarization dynamics in the time frame of 25–25,000 round trips (1 µs–1 ms) in terms of (d) Optical power of orthogonally polarized modes $I_x$ (dotted line) and $I_y$ (dashed line), total power $I = I_x + I_y$ (solid line); (e) Phase difference and degree of polarization; and (f) Normalized Stokes parameters at Poincaré sphere. Parameters: pump current $I_p$ = 240 mA, period $T$ = 38.9 ns, pulse width $T_p$ = 494 fs, output power $I \approx 0.25$ mW. (Reprinted with permission from Mou et al., 2013.)

As follows from Figure 1.13a, and Figure 1.7e, the spectrum indicates an interleaving of independent tightly bound states with phase shifts of $\pi/2$ and $\pi$. The SOPs of two interleaved BSs are slightly different and so we have superposition of the SOP switching with cyclic trajectory on the Poincaré sphere with the period of 14 round trips (Figure 1.13b).

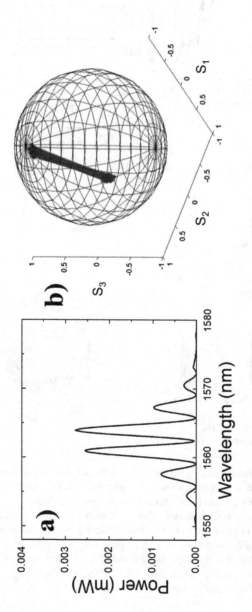

**FIGURE 1.10** Polarization dynamics of bound state soliton in the form of polarization switching between two orthogonal SOPs. (a) Output optical spectrum indicates π shift bound state with 370 fs pulse width and 1.5 ps pulse separation; (b) Stokes parameters on the Poincaré sphere. Each point in Figure 1.10b corresponds to a single laser pulse. (Adapted from Tsatourian et al., 2013a.)

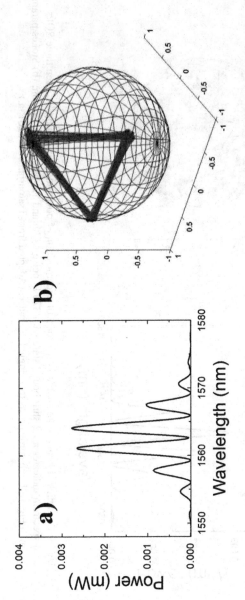

**FIGURE 1.11**  Polarization dynamics of bound state soliton in the form of polarization switching between three SOPs. (a) Output optical spectrum indicates π shift bound state with 370 fs pulse width and 1.5 ps pulse separation; (b) Stokes parameters on the Poincaré sphere. Each point in Figure 1.11b corresponds to a single laser pulse. (Adapted from Tsatourian et al., 2013a.)

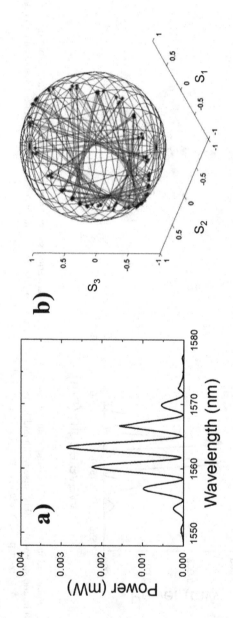

**FIGURE 1.12**  Polarization dynamics of bound state soliton in the form of superposition of polarization switching between three SOPs and SOP precession. (a) Output optical spectrum indicates $\pi/2$ shift bound state with 370 fs pulse width and 1.5 ps pulse separation; (b) Stokes parameters on the Poincaré sphere. Each point in Figure 1.12b corresponds to a single laser pulse. (Adapted from Tsatourian et al., 2013a.)

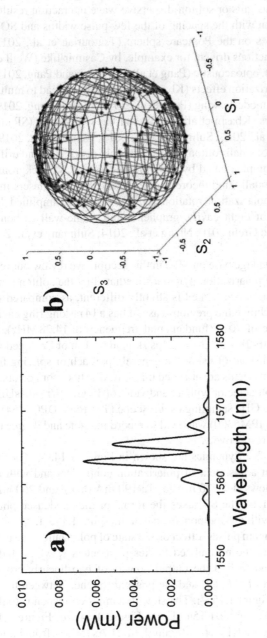

**FIGURE 1.13** Polarization dynamics of bound state soliton in the form of superposition popularization switching between two SOPs of two interleaved BSs and SOP precession. (a) Output optical spectrum indicates interleaved BSs with phase shifts of $\pi/2$ and $\pi$, 740 fs pulse width and 1.5 ps pulse separation; (b) Stokes parameters on the Poincaré sphere. Each point in Figure 1.13b corresponds to a single laser pulse. (Adapted from Tsatourian et al., 2013a.)

## 1.5 VECTOR SOLITON RAIN (EXPERIMENT)

It was demonstrated in previous sections that short-range interaction through the overlapping of solitons tails or soliton-dispersive wave interaction results in soliton bound states formation with the spacing of the few pulse widths and SOP evolution at different trajectories on the Poincaré sphere (Tsatourian et al., 2013a). Unlike this, long-range interactions driven, for example, by Casimir-like (Weill et al., 2016; Sulimany et al., 2018), optoacoustic (Pang et al., 2015; Liu and Pang, 2019; Sergeyev et al., 2021), and polarization effects (Kbashi et al., 2020) can lead to multi-pulsing in the form of harmonic mode-locking (Pang et al., 2015; Liu and Pang, 2019; Sergeyev et al., 2021), breathers (Kbashi et al., 2020), and the soliton rain (SR) (Chouli and Grelu 2010; Niang et al., 2014; Sulimany et al., 2018; Kbashi et al., 2019a).

The SR is a bunch of small soliton pulses randomly distributed and drifting nearby the main pulse and complemented by a continuous wave (cw) background. The SR was studied experimentally and theoretically for different fiber lasers mode-locked based on nonlinear polarization rotation (NPR), nonlinear amplified loop mirror (NALM), the figure of eight cavity, graphene, and single-wall carbon nanotubes (SWCNT) (Chouli and Grelu 2010; Niang et al., 2014; Sulimany et al., 2018; Kbashi et al., 2019a).

To extend the knowledge base on SRs, in this section, we review our recent experimental results on the polarization dynamics mediated by the soliton rain evolution in the laser cavity. The mode-locked is slightly different, as compared to the laser shown in Figure 1.1. Unlike the previous case, it has a 14 m-long ring cavity and so a photon roundtrip time of 70 ns (fundamental frequency of 14.28 MHz). The cavity comprises 13 m of SMF-28 with $\beta_2 = -22 \text{ps}^2/\text{km}$ and a 1 m of Er-doped fiber (EDF: Liekki Er80-8/125). Instead of using the general approach of splicing fibers in the ring cavity, all the components are attached through APC fiber connectors. The signal was detected by a photodetector with a bandwidth of 17 GHz (InGaAsUDP-15-IR-2 FC) connected to a 2.5 GHz sampling oscilloscope (Tektronix DPO7254). An in-line polarimeter (Thorlabs IPM5300) was used to record the state and degree of polarization (SOP and DOP), respectively.

The results on the fast dynamics are shown in Figure 1.14a–f. The results have been obtained without adjustment of polarization controllers and with a changing pump current J as follows: J = 220 mA (a, d); 190 mA (b, e); and 260 mA (c, f). As follows from Figure 1.14, the SR takes the forms of the condensed phase (Figure 1.14a, b) and a burst with cw component shown in Figure 1.14c, f.

To study the soliton rain pulses' effect on the state of polarization, we use IPM5300 polarimeter to measure the normalized Stokes parameters $s_1$, $s_2$, $s_3$, and degree of polarization (DOP) and to find the output powers of two linearly cross-polarized SOPs $I_x, I_y$, total power $I = I_x + I_y$ and the phase difference between them $\Delta\varphi$. The results are shown in Figure 1.15a–i. The slow dynamics corresponds to the cases of the fast dynamics, i.e., Figure 1.15a, d, g – Figure 1.14a, d, Figure 1.15b, e, h – Figure 1.14b, e, and Figure 1.15c, f, i – Figure 1.14c, f. As follows from Figure 1.15a, b,

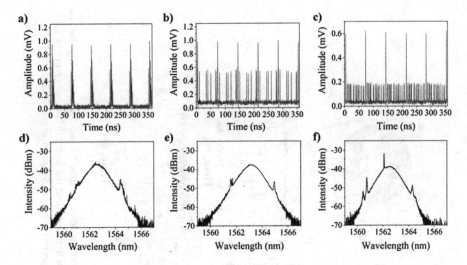

**FIGURE 1.14**  Fast dynamics. Oscillograms (a–c), and corresponding optical spectra (d–f). Parameters: Pump current J = 220 mA (a, d); 190 mA (b, e); 260 mA (c, f) (Reprinted with permission from Sergeyev et al., 2022).

spiral attractors can emerge in an almost isotropic cavity as a result of polarization symmetry breaking (Sergeyev, 2014; Sergeyev et al., 2014). In terms of the dynamics of the cross-polarized modes and the phase difference, the spiral attractor demonstrates the antiphase dynamics of the polarized SOPs the phase difference oscillations and switching between states of $\pi/2$ and $3\pi/2$ (Figure 1.15d,g). The high DOP of about 90% justifies that polarization evolution can be mapped with 1 ms resolution (Figure 1.15g). Emergence of small repulsing of the SR satellite pulses from the main pulse (Figure 1.14b) results in a small modification of polarization dynamics (Figure 1.15b, e, h). The dropped DOP to 80% is proof that dynamics is faster than 1 ms. With the increased number and the distance of the satellite pulses from the main pulse (Figure 1.14c, f), polarization attractor transforms from spiral to circle, as shown in Figure 1.15c. The output powers for two linearly cross-polarized SOPs are oscillating in antiphase whereas the phase difference dynamics is oscillating between states of $\pi/2$ and $3\pi/2$ (Figure 1.15f, i). The high DOP of 95% is an indication of slow dynamics with a resolution of 1 ms (Figure 1.15i).

The theoretical model for characterization of the vector soliton rain is found in Section 1.12.

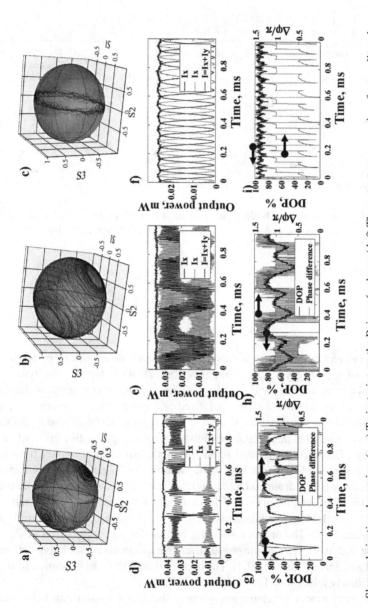

**FIGURE 1.15** Slow polarization dynamics: (a, c) Trajectories on the Poincaré sphere; (d, f) The output power vs time for two linearly cross-polarized SOPs $I_x$ (thin black line) and $I_y$ (grey line) and total power $I = I_x + I_y$ (thick black line); (g, f) DOP (black) and the phase difference (grey) vs time. Parameters: Pump current J = 220 mA (a, d, g); 190 mA (b, e, h); 260 mA (c, f, i). (Reprinted with permission from Sergeyev et al., 2022.)

## 1.6  VECTOR BRIGHT-DARK ROGUE WAVES (EXPERIMENT)

The extreme events (rogue waves, RWs) can have an anomalously high amplitude and can emerge and disappear unpredictably. The RWs have initially observed in oceanography (Kharif et al., 2009) and further in such fields as financial markets (Yan, 2010), nonlinear optics and laser physics (Solli et al., 2007; Baronio et al., 2012; Onorato et al., 2013; Chen et al., 2014; Dudley et al., 2014; Akhmediev et al., 2016; Kbashi et al., 2018; Sergeyev et al., 2018). To be considered as RWs, extreme events should have probability higher than the probabilities for Gaussian or Rayleigh distributions and amplitudes more than twice as large as the significant wave height (SWH). The SWH was initially defined as the mean amplitude of the highest third of the waves (Kharif et al., 2009) and, at present, it is more common to use SWH definition as amplitude equals to the four times of the standard deviation of the amplitude's variations (Onorato et al., 2013).

The scarcity of RWs and the inability to perform full-scale experiments in real-world scenarios are the major obstacles for developing techniques for RWs prediction and mitigation. Given mode-locked fiber lasers' (MLFLs') ability to generate pico- and femtosecond pulses with MHz repetition rates, more data on rogue waves in the short run (compared to the time scale of RWs in other systems, such as in the ocean and financial market (Kharif et al., 2009; Yan, 2010) can be collected under laboratory-controlled conditions. Previously, it has been found that RW can emerge in mode-locked lasers in the form of the soliton rain at the time scale of a roundtrip time (Solli et al., 2007; Onorato et al., 2013; Dudley et al., 2014; Akhmediev et al., 2016; Kbashi et al., 2018; Sergeyev et al., 2018).

All of the above experimental observations report the existence of either bright or dark rogue waves. The co-existence of the bright-dark rogue waves (BDRWs) has been predicted theoretically using coupled nonlinear Schrödinger equation (NLSE) systems (Baronio et al., 2012; Chen et al., 2014), but has never been observed experimentally in optics. In this section, we review our experimental results on a new mechanism of bright-dark rogue wave caused by desynchronization of the linear states of polarizations (SOPs) (Kbashi et al., 2018; Sergeyev et al., 2018).

Unlike the previous design, an Er-doped fiber laser mode-locked by CNT comprises a 1.1 m-long Er-doped fiber (EDF) with absorption of 80 dB/m at 1,530 nm and the group velocity dispersion (GVD) of +59 ps$^2$/km. A standard 70:30 output coupler (OUTPUT C) redirected 30% of the laser light out of the cavity. In addition, the laser cavity has 1.22 m of OFS980 fiber and 4.4 m of SMF 28 fiber with the GVD of −0.04 ps$^2$/nm. The POC1 and POC2 have been adjusted to find conditions for the emergence of RWs. To find the probability distribution histograms, the output voltage V for the oscilloscope and output power for the polarimeter is normalized as $V_n = (V-\text{median}(V))/\sigma(V)$, and so the RW criterion takes the form of $V_n > 8$.

To eliminate the soliton-soliton and soliton-dissipative wave interactions at the fast (roundtrip time) scale, we decreased the pump power $P = 18.4$ mW and tuned POC1 and POC2 to suppress soliton rain. As a result, we observed the dark-bright rogue waves as shown in Figure 1.16a–d. Unlike the previous cases shown in Figures 1.14 and 1.15, the output power is randomly changing from pulse to pulse (Figure 1.16a, d)) and is satisfying the BDRWs' criteria (Figure 1.16c). The slow polarization

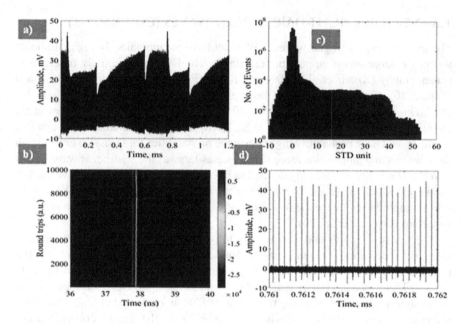

**FIGURE 1.16** Laser dynamics demonstrating the emergence of the bright-dark rogue waves at pump power of 18.4 mW. (a) Oscilloscope traces (32 GHz resolution); (b) Spatio-temporal dynamics (roundtrip time vs a number of round trips); (c) PDF histogram; (d) Part of the oscillogram demonstrating the absence of the soliton rain. (Reprinted with permission from Kbashi et al., 2018.)

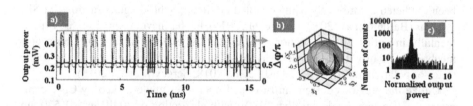

**FIGURE 1.17** Polarization laser dynamics at the time scale of tens of thousands of round trips. (a)–(c) polarization measurements (1 μs resolution, i.e., averaging over approximately 33 round trips, 16 slices with 1,024 points per slice): (a) The output power vs time (blue) and the phase differences (red); and (b) Trajectories in normalized Poincaré sphere. (c) Probability distribution histogram for the total output power $I = I_x + I_y$. The output power $I$ is normalized as shown in Figure 1.1. Parameters: (a)–(c) $P = 18.4$ mW. (Adapted from Kbashi et al., 2018.)

dynamics measured by IP5300 is shown in Figure 1.17a–c. As follows from Figure 1.17a, b, the anomalous spikes in the output power satisfy the RWs' criteria and are accompanied by transitions between orthogonally polarized SOPs, i.e., the phase difference jumps in $\pi$ (Figure 1.17a, b). Thus, the experimental data demonstrate BDRWs' appearance at the fast (roundtrip scale) and slow (tens of thousands of roundtrips) time scales for $P = 18.4$ mW.

The mechanism of BDRWs' emergence based on the desynchronization of orthogonally polarized SOPs is justified in Section 1.13 with the help of a new vector model which is different from the previously developed models based on coupled Schrödinger or Ginzburg-Landau equations (Kbashi et al., 2018; Sergeyev et al., 2018).

## 1.7  VECTOR RESONANCE MULTIMODE INSTABILITY (EXPERIMENT)

Modulation instability (MI) is a mechanism driving the emergence of spatial and temporal patterns in fluids, granular media, plasma, nonlinear optics and lasers (Faraday, 1831; Benjamin and Feir, 1967; Szwaj et al., 1998; Onorato et al., 2009; Zakharov and Ostrovsky, 2009; Agrawal, 2013; Turitsyna et al., 2013; Tlidi et al., 2014; Perego et al., 2016) One of the MI cases, the Benjamin-Feir instability (BFI), is related to the origin of the structures with the wave numbers $k$ and $-k$ due and their synchronization with homogeneous mode of $k = 0$ trough nonlinearity (Benjamin and Feir, 1967; Tlidi et al., 2014). For Faraday instability, emerging spatial structures are result of an external uniform modulation (Faraday, 1831; Tlidi et al., 2014). The other type of MI, namely dissipative parametric instability (DPI) (Perego et al., 2016), is driven by periodic antiphase modulation of spectrally dependent losses towards formation of stable one- and two-dimensional patterns. Unlike MI, the main feature of multimode Risken-Nummedal-Graham-Haken (RNGH) instability is the presence of the second lasing threshold exceeding the first in nine times in terms of the pump power (Graham and Haken, 1968; Risken and Nummedal, 1968). For the pump power above the second threshold, excitation of the large number of the longitudinal spatial modes leads to generating the pulse train with period of the cavity roundtrip time. Since 1968, when RNGH instability was discovered, it was found that a new second lasing threshold is close to the first threshold for Er-doped fiber lasers (Fontana et al., 1995; Pessina et al., 1997; Pessina et al., 1999; Voigt et al., 2004; Lugiato et al., 2015). Also, it was recently found that with accounting for the vector nature of the fiber laser dynamics a new type of RNGH instability, vector resonance multimode instability (VRMI), can emerge (Sergeyev et al., 2017). In this section, we review our recent results on VRMI (Sergeyev et al., 2017). The increased in-cavity birefringence strength causes spatial SOP modulation of the in-cavity lasing field (with a period of the beat length) and emergence of the additional satellite frequencies with the frequency splitting proportional to the birefringence strength. When the splitting is approaching the frequency difference between the longitudinal modes, parametric resonance results in longitudinal modes synchronization and locking similar to the injection locking (Cundiff and Ye, 2003). In the experiments, the special laser configuration exclude mode locking based on nonlinear polarization rotation (Lee and Agrawal, 2010).

Unlike the previous laser setups, the schematic in Figure 1.18a includes 1m of Er-doped fiber (Liekki Er80-8/125) and 614 m of single mode fiber SMF-28. The 80/20 fiber coupler was used to redirect the part of the signal outside the cavity. The cavity was pumped via a 1480/1550 WDM by using a 1,480 nm laser diode (FOL14xx series) with an in-built isolator. The first lasing threshold for the continuous wave (CW) regime was found for 16 mW pump power whereas the second threshold of the

a)   b)

c)

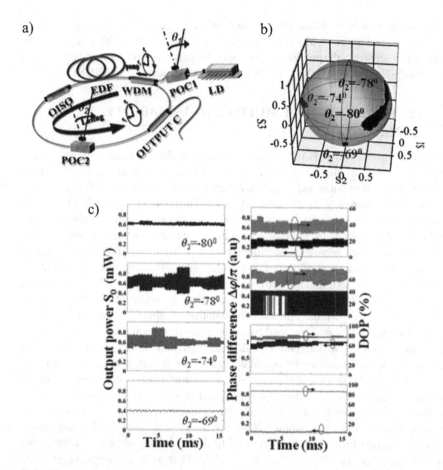

**FIGURE 1.18** (a) Erbium doped fiber laser. EDF: Erbium-doped fiber; LD: 1,480 nm laser diode for pump; POC1 and POC2: Polarization controllers, OISO: Optical isolator; WDM: Wavelength division multiplexer (WDM), OUTPUT C: 80:20 output coupler; (b) The map of the states of polarization on the Poincaré sphere (b) Output power ($S_0$) (c left) and corresponding phase difference between linearly polarized modes and DOP (c right) for different setting of the POC2: $\theta_2 = -80°$, $-78°$; $-74°$; $-69°$. (Adapted from Sergeyev et al., 2017.)

multimode instability was for 18 mW. The angles of the orientations of the paddles of POC1 and POC2 ($\theta_1$ and $\theta_2$) were measured from the vertical position. The $\theta_1 = -59°$ whereas $\theta_2$ was set at four different positions $\theta_2 = -80°$, $-78°$, $-74°$, $-69°$.

The results for polarization dynamics are shown in Figure 1.18b, c. As follows from the Figure 1.18b, decreasing the size of the spot by adjusting $\theta_2$ indicates that N-fold beat length is converging to the cavity length. As follows from Figure 1.18c, small oscillations of the output power $S_0$ and the phase difference $\Delta\varphi$ and small DOP of 40% justifies that the laser dynamics is faster than 1 μs for POC2 setting at $\theta_2 = -80°$. The further tuning the POC2 from $\theta_2 = -78°$ to $\theta_2 = -69°$ demonstrates the constant outputs and high (over 80%) DOP results in stable mode and SOP locking caused by matching N-fold beat length to the cavity length.

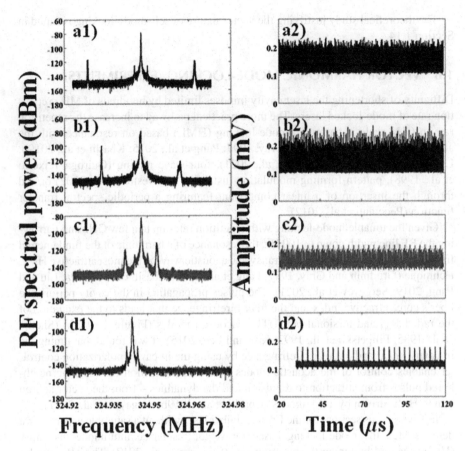

**FIGURE 1.19** The RF spectrum (a1–d1) and corresponding oscillograms (a2–d2) for different setting of the POC2: (a1, a2) $\theta_2 = -80°$; (b1, b2) $\theta_2 = -78°$; (c1, c2) $\theta_2 = -74°$; (d1, d2) $\theta_2 = -69°$. (Adapted from Sergeyev et al., 2017.)

Adjustment of the in-cavity polarization controller was resolved better in the case of 1,000th harmonic and so the radiofrequency (RF) spectrum evolution was recorded for frequencies of around 325.2 MHz (Figure 1.19a1–d1). The RF spectrum in Figure 1.19a1–d1 has three types of peaks, including the 1,000th harmonic (central peak), two satellites adjustable with the help of the POC2 and two close peaks closely position of which is independent on POC2 adjustment. The origin of the satellite frequencies is discussed further in the theoretical part. As follows from Figure 1.19a2–d2, the adjustment of POC2 results in regime stabilization similar to mode-locking when satellite frequencies match the main line (Figure 1.19d2). For RHGM instability, two-mode operation is oscillations close the harmonic with the photon round trip time period (Fontana et al., 1995; Pessina et al., 1997; Pessina et al., 1999; Voigt et al., 2004; Lugiato et al., 2015). Unlike this, as follows form Figure 1.19d2, the pulse width is of 40 ns vs the round-trip time of 3μs. So, many longitudinal modes are phase synchronized (Sergeyev et al., 2017).

The theoretical study justifying the vector nature of self-mode-locking is found in Section 1.14.

## 1.8 VECTOR HARMONIC MODE-LOCKING (EXPERIMENTS)

Difficulty of shortening the laser cavity imposes limited by hundreds of MHz repetition rate of mode-locked lasers. The more practical pathway to increase the repetition rate to GHz scale is harmonic mode-locking (HML) based on resonance with the acoustic phonons (Grudinin and Gray, 1997; Pang et al., 2015; Kbashi et al., 2019b; Liu and Pang, 2019; Sergeyev et al., 2021), four-wave mixing (Quiroga-Teixeiro et al., 1998), pattern-forming modulation instability (Sylvestre et al., 2002) or/and through the insertion of a linear component featuring a periodic spectral transfer function (Peccianti et al., 2012).

Given the tunable mode-locking with repetition rates up to a few GHz and narrowing the RF line width down to 100 Hz, the resonance of a harmonic of the fundamental frequency with the frequency of a transverse acoustic wave is the most attractive HML technique (Grudinin and Gray, 1997; Pang et al., 2015; Kbashi et al., 2019a; Liu and Pang, 2019; Sergeyev et al., 2021). The pulses propagating in the cavity perturb the fiber's core refractive index and the fiber birefringence that leads to the excitation of the radial $R_{0m}$ and torsional-radial $TR_{2m}$ acoustic modes (Figure 1.20a, b; (Shelby et al., 1985; Pilipetskii et al., 1993; Kim and Lee, 2015). It was found that tuning the in-cavity linear and circular birefringence by using the in-cavity polarization controller enables control of the acoustic modes mediated interaction between the neighbored pulses from attraction to repulsion and the dynamics – from the vector soliton rain to HML driven by $TR_{2m}$ modes (Kbashi et al., 2019b; Sergeyev et al., 2021).

In the previous section, for the Er-doped fiber laser without a saturable absorber, we demonstrated the mode-locking based on vector resonance multimode instability (VRMI) caused by tuning the birefringence (Sergeyev et al., 2017). The $TR_{2m}$ modes induces weak oscillations of the fiber birefringence vector orientation (Shelby et al., 1985; Pilipetskii et al., 1993; Kim and Lee, 2015) and so there is a challenging task of revealing the interplay between VRMI and $TR_{2m}$ acoustic modes-based perturbation towards HML. In this section, we review our recent experimental results on novel vector HML mechanism caused by interplay of VRMI and $TR_{2m}$ (Sergeyev et al., 2021). The resonance occurred for the 24th, 38th, and 45th harmonics and resulted in linewidth narrowing below the values reported the other authors (Grudinin and Gray, 1997).

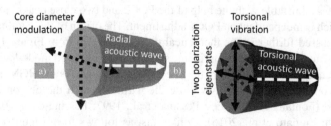

**FIGURE 1.20** Acoustic modes in optical fiber core: (a) Radial mode $R_{0m}$; (b) Torsional-radial mode $TR_{2m}$. (Adapted from Sergeyev et al., 2021.)

The setup is shown in Figure 1.21a. The cavity includes 15.8 m of SMF28 and 75 cm of Liekki Er80-8/125 Er-doped fiber with the anomalous dispersion of −20 fs²/mm. The total length of the cavity is 17 m. The pump laser diode (FOL14xx series with isolator) with the maximum power of 250 mW is used to pump the laser cavity via WDM coupler. A polarization controller POC1 and an optical isolator for 1,560 nm (to improve the laser diode stability) is located between the diode output and the WDM. The output coupler (OUTPUT C) 80:20 redirects the light out of the cavity. After installation of an isolator with 51 dB attenuation, the laser was successfully mode locked. The lasing threshold was measured as 36 mW of the pump power based on linear extrapolation of the signal versus pump power curve (Figure 1.21b). To characterize the polarization laser dynamics, a IPM5300 polarimeter is used. Given the absence of a polarizer and the presence of only one polarization controller inside the laser cavity, and low pump powers (less than 200 mW), mode-locking through nonlinear polarization rotation is excluded.

The graphs "or the output power versus pump power, the emission" spectrum, and the pulse train are shown in Figure 1.21b–d. For the pump power above 48 mW, mode-locked pulses emerges with the fundamental repetition rate of 12.21 MHz (Figure 1.21d), RF linewidth 370 Hz (INSET of Figure 1.21b). The transient time for stabilization of this regime varies from a fraction of a second to a few minutes. The pulse width of 20 ps (INSET of Figure 1.21d) cannot be measured with an autocorrelator and so, to estimate the pulse parameters, we used an ultrafast photodetector XPDV232OR with a bandwidth of 50 GHz and DSO-X93204A oscilloscope with a bandwidth of 32GHz. The pulse width of 20 ps was obtained using the oscilloscope trace and the interpolation software supplied by Agilent that gave us the effective resolution of 781 fs/point (Sergeyev et al., 2021). The low signal-to-noise ratio (SNR) of 6 dB (Figure 1.21d) reveals the partial mode-locking. The experimental results demonstrate the stable patterns at the fundamental frequency of 12.21 MHz and its high-order harmonics at frequencies of 293.16 MHz, 464.17 MHz, and 549.7 MHz (Table 1.1).

It has been demonstrated by many authors, that the excitation of oscillations at such frequencies is caused by the resonance structure of the spectrum of acoustic phonons excited through the electrostriction effect (Grudinin and Gray, 1997; Pang et al., 2015; Kbashi et al., 2019b; Liu and Pang, 2019; Sergeyev et al., 2021).

The dynamics of the HML at 293.16 MHz is shown in Figures 1.22a–d. A part of the RF spectra is in Figure 1.22a. The lines "A," "B" and "C" are related to the 23rd, 24th, and 25th harmonics of the fundamental frequency. We adjusted birefringence by turning the knob of the POC2 and fixed the pump power at 160 mW to clarify evolution of the satellite lines caused by linear and circular birefringence. When the angle of the knob was set between 18 positions, the satellites of the lines "A" and "C" were moving closer to the line "B" as shown in Figure 1.22a. To demonstrate the linewidth compression, we recorded temporal traces and RF spectra for the last four steps 15–18 (Figures 1.22b, c). For position 15 in Figures 1.22b, c, the distance between the satellites is slightly less than 3 MHz and the satellites disappear. The RF spectral line corresponds to the fundamental comb frequency with SNR changed from 6 dB to 30 dB for pump power increase from 48 mW to 160 mW. In position 16, the distance between satellites decreases, optical noise spectrum demonstrates a

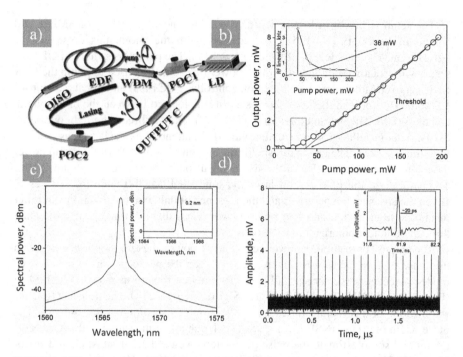

**FIGURE 1.21** Operation of the laser at the fundamental frequency (a) Erbium-doped fiber laser. EDF, Erbium fiber; LD, 1480 nm laser diode for the pump; POC1 and POC2, Polarization controllers; OISO, Optical isolator; WDM, Wavelength division multiplexer (WDM); OUTPUT C, 80:20 output coupler; (b) Average laser output power versus pump power; INSET: the RF linewidth versus pump power (370 Hz at 220 mW pump power). The rectangle indicates the interval where unstable mode-locking patterns have been observed; (c) The optical spectrum; inset: the same spectra plotted using a linear scale: 0.2 nm is a bandwidth at 3 dB level; (d) The train of pulses at the fundamental frequency, INSET: time-resolved pulse. (Reprinted with permission from Sergeyev et al., 2021.)

## TABLE 1.1
### Frequencies Observed in the Experiments

| Frequency, MHz | RF Peak Width, Hz | Temporal Jitter, ppm[3] | Long-Term Drift |
|---|---|---|---|
| 12,21 | [210, 370, 530][1,2] | 40 | Yes |
| 97.7 | Unstable | Unstable | - |
| 207.6 | Unstable | Unstable | - |
| 293.16 | [9, 38, 155][1] | 1.4 | Yes |
| 464.17 | [22, 38, 150][1] | 0.9 | Yes |
| 549.7 | [1, 13, 97][1] | 0.5 | Yes |
| 842.5 | Unstable | Unstable | - |
| 903.5 | Unstable | Unstable | - |

Note:
1 Asymmetric interval of confidence 0.95 [min, mean, max]
2 At pump power of 220 mW
3 Parts per million with respect to the main value of frequency. The jitter has been quantified using ARIMA (0, 1, 0) (random walk with drift) model with the interval of confidence 0.95.

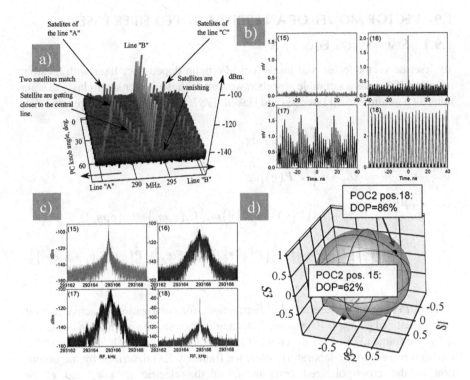

**FIGURE 1.22** Acousto-optical polarization-dependent locking of a high harmonics (a) RF comb showing 24th harmonic along with satellites of 23rd, 24th, and 25th harmonics tuning with the help of in-cavity polarization controller POC2; (b) Emergence of the 293.16 MHz pulse train for the positions 15–18 of the POC2; (c) Evolution of the RF spectrum of the 293.16 MHz line for the positions 15–18 of the POC2; (d) The output SOPs for the POC2 positions 15 and 18 (measurements resolution is 1 μs). (Reprinted with permission from Sergeyev et al., 2021.)

periodic pattern, and the RF spectrum becomes broader. After the knob of POC2 has turned to position 17, the oscilloscope traces (Figure 1.22b, shows oscillations at 293.16 MHz with the period close to 20 ns (50 MHz) and The RF spectrum has multiple peaks.

Finally, for the knob position 18, the modulation disappears, and the regular oscillations pattern at the frequency of 293.16 MHz emerges. The RF spectrum shows a narrow resonance line with 60 dB SNR and the noise level of −120 dB as shown in Figure 1.22d. The locked SOP for POC2 positions 15 and 18 corresponds to the self-oscillation at the fundamental frequency for position 15 and HML for position 18. The tuning POC2 from position 15 to position 18 changes the linear and circular birefringence in the cavity due to induced fiber squeezing and twist (Collett, 2003). As shown in Figure 1.22d, increasing DOP from 62% to 86% indicates more stable operation for 18th position as compared to the 15th position. By adjusting POC2. HML at different acoustic frequencies is observed as shown in Table 1.1. In Section 1.15, we review our recent theoretical results on novel vector HML mechanism caused by interplay of VRMI and $TR_{2m}$ (Sergeyev et al., 2021).

## 1.9  VECTOR MODEL OF AN ERBIUM-DOPED FIBER LASER

### 1.9.1  Semiclassical Equations

To describe vector features of mode-locked erbium doped fiber laser with a carbon nanotube (CNT) as a saturable absorber, we start with the vector semi-classical equations for a unidirectional laser (Fu and Haken, 1987; Sergeyev, 1996; Sergeyev, 1999):

$$\frac{\partial E_x}{\partial t} + c\frac{\partial E_x}{\partial z} = -kE_x + ik\int\left(e_x P(g)\right)dg,$$

$$\frac{\partial E_y}{\partial t} + c\frac{\partial E_y}{\partial z} = -kE_y + ik\int\left(e_y P(g)\right)dg,$$

$$\frac{\partial P(g)}{\partial t} = \left(-\gamma_p + i\Delta_0\right)P(g) - i\gamma_p D(g)m_e^*\left(E_x\left(e_x m_e\right) + E_y\left(e_y m_e\right)\right),$$

$$\frac{\partial D(g)}{\partial t} = \gamma_d\left(D_0 - D(g) + \frac{i}{4}\left(P(g)*\left[E_x e_x + E_y e_y\right] - P(g)\left[E_x^* e_x + E_y^* e_y\right]\right)\right).$$

$$(1.4)$$

Here $P(g)$ and $D(g)$ are angular distributions for polarization of active medium and normalized gain, $g = (\theta,\varphi,\psi)$ are Euler angles showing the orientation of the local reference frame $(X',Y',Z')$ connected to the orientation of the dipole moments of $Er^{3+}$ ion with respect to the laboratory reference frame $(X,Y,Z)$ described by the orientation of the cross-polarized components of the electric field $e_x$ and $e_y$. So, $\int...dg = (1/8\pi)^2\int_0^{2\pi}\int_0^{\pi}...\sin d\theta d\phi d\psi$ (Varshalovich et al., 1988), $E = E_x e_x + E_y e_y$ is a lasing electric field, $m_e$ is a unit vector for the dipole moment of the transition with emission (Sergeyev, 1996; Sergeyev, 1999), $D_0$ is the normalized parameter for the pumping light power, $\Delta_0$ is detuning of the lasing wavelength from to the maximum of the gain spectrum, the vector $m_e^*$ is a vector with complex conjugation of the vector $m_e$; $k$, $\gamma_p$, and $\gamma_d$ are the relaxation rates for photons in the cavity, medium polarization and gain. Given the relaxation rate of the medium polarization in erbium doped silica matrix $\gamma_p = 4.75 \times 10^{14}$ s$^{-1}$ >> $\gamma_d$, $k$ ($\gamma_d = 100$ s$^{-1}$, k = $10^7 - 10^8$ s$^{-1}$) (Williams and Roy, 1996), we can use the following simplification $\partial P(g)/\partial t = 0$ and so Equation (1.4) takes the following form:

$$\frac{\partial E_x}{\partial t} + c\frac{\partial E_x}{\partial z} = -kE_x + \frac{(1+i\Delta)}{1+\Delta^2}\left(D_{xx}E_x + D_{xy}E_y\right),$$

$$\frac{\partial E_y}{\partial t} + c\frac{\partial E_y}{\partial z} = -kE_y + \frac{(1+i\Delta)}{1+\Delta^2}\left(D_{yx}E_x + D_{yy}E_y\right),$$

$$\frac{\partial D(g)}{\partial t} = \gamma_d\left(D_0 - D(g) - \frac{D(g)}{2}R\left(E_x,E_y,g\right)\right),$$

$$(1.5)$$

$$R\left(E_x,E_y,g\right) = \frac{1}{1+\Delta^2}\begin{bmatrix}\left|E_x\right|^2\left|e_x m_e\right|^2 + \left|E_y\right|^2\left|e_y m_e\right|^2 + \\ E_x E_y^*\left(e_x m_e\right)\left(e_y m_e^*\right) + E_y E_x^*\left(e_y m_e\right)\left(e_x m_e^*\right)\end{bmatrix}.$$

Here $\Delta = \Delta_0/\gamma_p$ and

$$D_{xx} = k\int D(g)\left|e_x m_e\right|^2 dg, \quad D_{yy} = k\int D(g)\left|e_y m_e\right|^2 dg,$$

$$D_{xy} = k\int D(g)\left(e_y m_e\right)\left(e_x m_e^*\right)dg, \quad D_{yx} = k\int n(g)\left(e_x m_e\right)\left(e_y m_e^*\right)dg \tag{1.6}$$

By adding a saturable absorber (single-wall carbon nanotubes, CNTs), fiber bire-fringence, Kerr nonlinearity, chromatic dispersion (Sergeyev, 2014; Sergeyev et al., 2014), absorption from the ground state at the lasing wavelength for Erbium ions (Desurvire, 1994) (Figure 1.23a), and SOP for pump wave, we can modify Equation (1.6) as follows (Sergeyev, 2014; Sergeyev et al., 2014):

$$\frac{\partial E_x}{\partial z} = i\beta E_x - \eta\frac{\partial E_x}{\partial t} - i\beta_2\frac{\partial^2 E_x}{\partial t^2} + i\gamma\left(\left|E_x\right|^2 E_x + \frac{2}{3}\left|E_y\right|^2 E_x + \frac{1}{3}E_y^2 E_x^*\right) + D_{xx}E_x + D_{xy}E_y,$$

$$\frac{\partial E_y}{\partial z} = -i\beta E_y + \eta\frac{\partial E_y}{\partial t} - i\beta_2\frac{\partial^2 E_y}{\partial t^2} + i\gamma\left(\left|E_y\right|^2 E_y + \frac{2}{3}\left|E_x\right|^2 E_y + \frac{1}{3}E_x^2 E_y^*\right)$$
$$+ D_{yx}E_x + D_{yy}E_y,$$

$$D_{xx} = \frac{\alpha_1}{2}\frac{(1-i\Delta)}{1+\Delta^2}\left[\chi\int n(g)\left|e_x m_e\right|^2 dg - 1\right] - \alpha_2\int N(g)\left|e_x \mu_\alpha\right|^2 dg - \alpha_4,$$

$$D_{xy} = \frac{\alpha_1 \chi}{2}\frac{(1-i\Delta)}{1+\Delta^2}\int n(g)\left(e_y m_e\right)\left(e_x m_e^*\right)dg - \alpha_2\int N(g)\left(e_y \mu_\alpha\right)\left(e_x \mu_\alpha^*\right)dg,$$

$$D_{yx} = \frac{\alpha_1 \chi}{2}\frac{(1-i\Delta)}{1+\Delta^2}\int n(g)\left(e_x m_e\right)\left(e_y m_e^*\right)dg - \alpha_2\int N(g)\left(e_x \mu_\alpha\right)\left(e_y \mu_\alpha^*\right)dg,$$

$$D_{yy} = \frac{\alpha_1}{2}\frac{(1-i\Delta)}{1+\Delta^2}\left[\chi\int n(g)\left|e_y m_e\right|^2 dg - 1\right] - \alpha_2\int N(g)\left|e_y \mu_\alpha\right|^2 dg - \alpha_{4r}$$

$$\frac{\partial n(g)}{\partial t} = \gamma_d\left(\frac{I_p}{I_{ps}}\left(1-n(g)\right)\left|e_p m_\alpha\right|^2 - n(g) - \left(\chi n(g) - 1\right)R_{E_r}\left(E_x, E_y, g\right)\right),$$

$$N(g) = \frac{1}{1+\alpha_3 R_{CNT}\left(E_x, E_y, g\right)\left(1+\Delta^2\right)}$$

$$R_{E_r}\left(E_x, E_y, g\right) = \frac{1}{1+\Delta^2}\left[\frac{\left|E_x\right|^2}{I_{ss}}\left|e_x m_e\right|^2 + \frac{\left|E_y\right|^2}{I_{ss}}\left|e_y m_e\right|^2 + \frac{E_x E_y^*}{I_{ss}}\left(e_x m_e\right)\left(e_y m_e^*\right)\right.$$
$$\left. + \frac{E_y E_x^*}{I_{ss}}\left(e_y m_e\right)\left(e_x m_e^*\right)\right],$$

$$R_{CNT}\left(E_x, E_y, g\right) = \frac{1}{1+\Delta^2}\left[\begin{array}{c}\dfrac{\left|E_x\right|^2}{I_{ss}}\left|e_x \mu_\alpha\right|^2 + \dfrac{\left|E_y\right|^2}{I_{ss}}\left|e_y \mu_\alpha\right|^2 \\ + \dfrac{E_x E_y^*}{I_{ss}}\left(e_x \mu_\alpha\right)\left(e_y \mu_\alpha^*\right) + \dfrac{E_y E_x^*}{I_{ss}}\left(e_y \mu_\alpha\right)\left(e_x \mu_\alpha^*\right)\end{array}\right].$$

$$\tag{1.7}$$

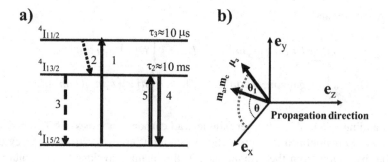

**FIGURE 1.23**  (a) Diagram of energy levels and transitions in Er³⁺ ion: 1, Pump absorption at 980 nm; 2, Non-radiative phonon-assisted transitions to the first excited level; 3, Fluorescence at 1,550 nm; 4, and 5, Stimulated emission and absorption at 1,550 nm; (b) The orientation of absorption and emission dipole moments for erbium doped silica ($m_a$, $m_e$) and the absorption dipole moment for CNT $\mu_a$. (Adapted from Sergeyev, 2014.)

Here $n(g)$ and $N(g)$ are the angular distributions of the erbium ions at the first excited level and CNT in the ground state, $\alpha_2$ is the CNT absorption at the lasing wavelength, $\alpha_3$ is the ratio of saturation powers for CNT and EDF, $\alpha_4$ represents the normalized losses, $\beta$ is the birefringence strength ($2\beta = 2\pi/L_b$, $L_b$ is the beat length), $m_a$ and $\mu_a$ are unit vectors along the dipole moment of the transition with absorption for erbium ions and CNT, $V_g$ is the group velocity, $\eta = \beta\lambda/(2\pi c)$ is the inverse group velocity difference between the polarization modes, $\alpha_1 = \sigma_a \Gamma_L \rho$ is the EDF absorption at the lasing wavelength, $I_{ps} = \gamma_d A h v_p \big/ \big(\sigma_a^{(p)}\Gamma_p\big)$, $I_{ss} = \gamma_d A h v_s \big/ \big(\sigma_a^{(L)}\Gamma_L\big)$ are saturation powers for pump and lasing ($h$ is Planck's constant, $v_p$, $v_s$ are pump and lasing frequencies), $\chi = \big(\sigma_a^{(L)} + \sigma_e^{(L)}\big)\big/\sigma_a^{(L)}$, $\sigma_{a(e)}^{(L)}$, $\sigma_a^{(p)}$ are absorption and emission cross-sections at the lasing wavelength and absorption cross-section at the pump wavelength, $\Gamma_L$ and $\Gamma_p$ are the confinement factors of the EDF fiber at the lasing and pump wavelengths, $\rho$ is the concentration of erbium ions, $A$ is the fiber core cross-section area.

### 1.9.2  REDUCING THE COMPLEXITY OF THE SEMICLASSICAL MODEL

To simplify the description of the polarized lasing field interaction with the gain medium in Equation (1.7), we use an approximation introduced by Zeghlache and Boulnois (1995) and justified by Leners and Stéphan (1995), i.e., we suggest that dipole moments ($m_a$, $m_e$) and the dipole moment $\mu_a$ are located in the plane defined by the orthogonal components of the lasing field $e_x$ and $e_y$ (Figure 1.23b). In addition, we use the property of Er ions $m_a = m_e$ and consider an elliptically polarized pump $e_p = (e_x + i\delta e_y)\big/\sqrt{1+\delta^2}$ (here $\delta$ is the ellipticity of the pump wave) as follows (Sergeyev, 2014; Sergeyev et al., 2014)

$$\left(m_e e_x\right) = \cos(\theta), \left(m_e e_y\right) = \sin(\theta), \left(m_\alpha e_p\right)^2 = \frac{\cos(\theta)^2 + \delta^2 \sin(\theta)^2}{1+\delta^2}, \quad (1.8)$$
$$\left(\mu_\alpha e_x\right) = \cos(\theta_1), \left(\mu_\alpha e_y\right) = \sin(\theta_1)$$

So, the angular distributions $n(g)$ now depends only on $\theta$ and can be presented by a Fourier series (Zeghlache and Boulnois, 1995; Sergeyev, 2014; Sergeyev, 2014):

$$n(\theta) = \frac{n_0}{2} \sum_{k=1}^{\infty} n_{1k} \cos(k\theta) + \sum_{k=1}^{\infty} n_{2k} \sin(k\theta). \quad (1.9)$$

Substituting Equation (1.9) to Equation (1.4), we find a complete set of equations for $E_x$, $E_y$, $n_0$, $n_{12}$, $n_{22}$ (Sergeyev, 2014; Sergeyev et al., 2014):

$$\frac{\partial E_x}{\partial z} = i\beta E_x - \eta \frac{\partial E_x}{\partial t} - i\beta_2 \frac{\partial^2 E_x}{\partial t^2}$$
$$+ i\gamma \left( |E_x|^2 E_x + \frac{2}{3}|E_y|^2 E_x + \frac{1}{3} E_y^2 E_x^* \right) + D_{xx} E_x + D_{xy} E_y,$$

$$\frac{\partial E_y}{\partial z} = i\beta E_y + \eta \frac{\partial E_y}{\partial t} - i\beta_2 \frac{\partial^2 E_y}{\partial t^2} + i\gamma \left( |E_y|^2 E_y + \frac{2}{3}|E_x|^2 E_y + \frac{1}{3} E_x^2 E_y^* \right)$$
$$+ D_{yx} E_x + D_{yy} E_y$$

$$D_{xx} = \left( \frac{\alpha_1(1-i\Delta)}{1+\Delta^2} I_{xx}\left(n_0, n_{12}, n_{22}\right) - J_{xx} - \alpha_4 L \right),$$

$$D_{xy} = D_{yx} = \left( \frac{\alpha_1(1-i\Delta)}{1+\Delta^2} \times I_{xy}\left(n_0, n_{12}, n_{22}\right) - J_{xy} \right),$$

$$D_{yy} = \left( \frac{\alpha_1(1-i\Delta)}{1+\Delta^2} I_{yy}\left(n_0, n_{12}, n_{22}\right) - J_{yy} - \alpha_4 L \right)$$

$$I_{xx}\left(n_0, n_{12}, n_{22}\right) = \left( \chi \frac{n_0}{2} - 1 \right) + \chi \frac{n_{12}}{2},$$

$$I_{yy}\left(n_0, n_{12}, n_{22}\right) = \left( \chi \frac{n_0}{2} - 1 \right) - \chi \frac{n_{12}}{2},$$

$$I_{xy}\left(n_0, n_{12}, n_{22}\right) = \chi \frac{n_{22}}{2},$$

$$J_{xx} = \alpha_2 \left( \frac{1}{2} - \alpha_3 \frac{1}{8} \left[ 3|E_x|^2 + |E_y|^2 \right] \right).$$

$$J_{yy} = \alpha_2 \left( \frac{1}{2} - \alpha_3 \frac{1}{8} \left[ |E_x|^2 + 3|E_y|^2 \right] \right),$$

$$J_{xy} = -\frac{\alpha_3 \alpha_2}{8} \left[ E_x E_y^* + c.c. \right],$$

$$\frac{dn_0}{dt} = \gamma_d \left[ I_d + 2R_{10} - \left(1 + \frac{I_p}{2} + \chi R_{10}\right) n_0 - \left(\chi R_{11} + \frac{I_p}{2} \frac{\left(1-\delta^2\right)}{\left(1+\delta^2\right)}\right) n_{12} - \chi n_{22} \times R_{12} \right],$$

$$\frac{dn_{12}}{dt} = \gamma_d \left[ \frac{\left(1-\delta^2\right) I_p}{\left(1+\delta^2\right)2} + R_{11} - \left(\frac{I_p}{2} + 1 + \chi R_{10}\right) n_{12} - \left(\frac{\left(1-\delta^2\right)}{\left(1+\delta^2\right)} \frac{I_p}{2} + \chi R_{11}\right) \frac{n_0}{2} \right],$$

$$\frac{dn_{22}}{dt} = \gamma_d \left[ R_{12} - \left(\frac{I_p}{2} + 1 + \chi R_{10}\right) n_{22} - \chi R_{12} \frac{n_0}{2} \right],$$

$$R_{10} = \frac{1}{2\left(1+\Delta^2\right)} \left( |E_x|^2 + |E_y|^2 \right),$$

$$R_{11} = \frac{1}{2\left(1+\Delta^2\right)} \left( |E_x|^2 - |E_y|^2 \right),$$

$$R_{12} = \frac{1}{2\left(1+\Delta^2\right)} \left[ E_x E_y^* + c.c. \right]$$

$$(1.10)$$

Here we use approximation $\frac{\alpha_3}{4}\left[3|E_x|^2 + |E_y|^2\right] << 1$ We apply the distributed forms for saturable absorption and losses in Equation (1.10) instead of lumped presentation to simplify consideration. Next, we use the approach of averaging over the pulse width to characterize the slow time scale dynamics. We introduce a new slow-time variable $t_s = z/\left(V_g t_R\right)$, where $t_r = L/V_g$ is the photon roundtrip time, $L$ is the cavity length) and assume an ansatz in the form (Sergeyev, 2014; Sergeyev et al., 2014):

$$E_x\left(t,t_s\right) = u\left(t_s\right) \sec h\left(t/T_p\right), E_y\left(t,t_s\right) = v\left(t_s\right) \sec h\left(t/T_p\right). \qquad (1.11)$$

Here $T_p$ is the pulse width. After substitution of Equation (1.11) into Equation (1.10) and averaging over the time $T_p << t << t_R$ we obtain the following (Sergeyev, 2014; Sergeyev et al., 2014):

$$\frac{du}{dt_s} = i\beta Lu + i\frac{\gamma L I_{ss}}{2}\left(|u|^2 u + \frac{2}{3}|v|^2 u + \frac{1}{3}v^2 u^*\right) + D_{xx}u + D_{xy}v,$$

$$\frac{dv}{dt_s} = -i\beta Lv + i\frac{\gamma L I_{ss}}{2}\left(|v|^2 v + \frac{2}{3}|u|^2 v + \frac{1}{3}u^2 v^*\right) + D_{xy}u + D_{yy}v,$$

$$\frac{dn_0}{dt_s} = \varepsilon\left[ I_p + 2R_{10} - \left(1 + \frac{I_p}{2} + \chi R_{10}\right) n_0 - \left(\chi R_{11} + \frac{I_p}{2}\frac{\left(1-\delta^2\right)}{\left(1+\delta^2\right)}\right) n_{12} - \chi n_{22} R_{12} \right],$$

$$\frac{dn_{12}}{dt_s} = \varepsilon\left[ \frac{\left(1-\delta^2\right)}{\left(1+\delta^2\right)}\frac{I_p}{2} + R_{11} - \left(\frac{I_p}{2} + 1 + \chi R_{10}\right) n_{12} - \left(\frac{\left(1-\delta^2\right)}{\left(1+\delta^2\right)}\frac{I_p}{2} + \chi R_{11}\right)\frac{n_0}{2} \right],$$

$$\frac{dn_{22}}{dt_s} = \varepsilon \left[ R_{12} - \left( \frac{I_p}{2} + 1 + \chi R_{10} \right) n_{22} - \chi R_{12} \frac{n_0}{2} \right],$$

$$R_{10} = \frac{1}{\left(1+\Delta^2\right)} \left( |u|^2 + |v|^2 \right), R_{11} = \frac{1}{\left(1+\Delta^2\right)} \left( |u|^2 - |v|^2 \right), R_{12} = \frac{1}{\left(1+\Delta^2\right)} \left( uv^* + vu^* \right),$$

$$(1.12)$$

Coefficients $D_{ij}$ can be found as follows:

$$D_{xx} = \frac{\alpha_1 L (1-i\Delta)}{1+\Delta^2} (f_1 + f_2) - \left( \frac{\alpha_2 L}{2} - \frac{2\alpha_2\alpha_3 L}{8\pi} k_1 \right) - \alpha_4 L,$$

$$D_{yy} = \frac{\alpha_1 L (1-i\Delta)}{1+\Delta^2} (f_1 - f_2) - \left( \frac{\alpha_2 L}{2} - \frac{\alpha_2\alpha_3 L}{4\pi} k_2 \right) - \alpha_4 L, \qquad (1.13)$$

$$D_{xy} = D_{yx} = \frac{\alpha_1 L (1-i\Delta)}{1+\Delta^2} f_3 - \frac{2\alpha_2\alpha_3 L}{8\pi} k_3,$$

where:

$$f_1 = \left( \chi \frac{n_0}{2} - 1 \right), f_2 = \chi \frac{n_{12}}{2}, f_3 = \chi \frac{n_{22}}{2},$$

$$(1.14)$$

$$k_1 = 3|u|^2 + |v|^2, k_2 = |u|^2 + 3|v|^2, k_3 = uv^* + vu^*.$$

Here $\varepsilon = t_R \gamma_d$ and $u$, $v$ are normalized to the saturation power $I_{ss}$ and $I_p$ is normalized to the saturation power $I_{ps}$. We have also neglected the inverse group velocity difference of the cross-polarized components x and y that corresponds to $\eta \approx 0$. For a cavity length $L_c = 7.8$ m, beat length $L_b = 5$ m, and $\lambda = 1.56$ μm the time delay between cross-polarized pulses over the length of the cavity can be found from the notations to Equation (1.9) as $T_d = 8$ fs. Given the $T_d \ll T_p = 600$ fs and the CNT relaxation time of 300 fs, the group velocity difference can be neglected in Equation (1.9). We have also used the following notations (Sergeyev, 2014; Sergeyev et al., 2014)

$$\int_{-T/T_p}^{T/T_p} \frac{\cos h(x)^2 - 2}{\cos h(x)^3} dx \rightarrow 0, \frac{\int_{-T/T_p}^{T/T_p} \sec h(x^3) dx}{\int_{-T/T_p}^{T/T_p} \sec h(x) dx} \approx \frac{1}{2}, \frac{\int_{-T/T_p}^{T/T_p} \sec h(x^3) dx}{\int_{-T/T_p}^{T/T_p} \sec h(x) dx} \approx \frac{2}{\pi}. \quad (1.15)$$

We have neglected the absorption dynamics in CNT that holds true for saturable absorber relaxation time $\tau_a$ is smaller than the pulse width $T_p$. In our experiments $\tau_a \sim 300$ fs and $T_p \sim 600$ fs and so the approximation of the fast saturable absorber is valid if we make change of variables $\alpha_2 \rightarrow \alpha_2(1-\exp(-T_p/\tau_a))$ for the case of $\alpha_2 \ll 1$. Though $Er^{3+}$ ion is usually described as a four-level system in Figure 1.23b, we reduce this model to a two-level one by excluding excited state absorption from $^4I_{11/2}$ and population of this level justified for pump powers $I_p$<200 mW, *viz.* for the case considered in our publications (Sergeyev, 2014; Sergeyev et al., 2014). In the case of a high-concentration Er-doped fiber, Sergeyev and co-workers demonstrated that migration assisted upconversion (MAUP) results in decreasing first excited level lifetime more than ten times (Sergeyev, 2003; Sergeyev et al., 2005; Sergeyev et al., 2006) and so decreasing the lifetime at the first excited level is a reliable approach for mimic MAUP (Sergeyev, 2014; Sergeyev et al., 2014). Slow MAUP dynamics in microseconds scale has no effect on pulse shape. Unlike previously used models based on either coupled nonlinear Schrödinger or Ginzburg-Landau equations, Equations (1.12–1.14) account for slow polarization dynamics.

To study the evolving SOPs of the vector solitons, we account for birefringence tuning by in-cavity polarization controller. First, we rewrite Equation (1.12) for $\Psi = (u,v)^T$ as

$$\psi(t_s+1) = B\exp(G)\psi(t_s),\qquad(1.16)$$

where

$$G = \begin{bmatrix} \int_{t_s}^{t_s+1} D_{xx}(t_s)dt_s & \int_{t_2}^{t_s+1} D_{xy}(t_s)dt_s \\ \int_{t_s}^{t_s+1} D_{xy}(t_s)dt_s & \int_{t_s}^{t_s+1} D_{yy}(t_s)dt_s \end{bmatrix}, B = \begin{bmatrix} \exp\left(\dfrac{i\pi L}{L_b}\right) & 0 \\ 0 & \exp\left(-\dfrac{i\pi L}{L_b}\right) \end{bmatrix}.\qquad(1.17)$$

The presence of the in-cavity polarization controller modifies Equation (1.16):

$$\psi(t_s+1) = TB\exp(G)\psi(t_s),\qquad(1.18)$$

Where $T$ is the transfer matrix of POC (Heismann, 1994)

$$T = \begin{bmatrix} A+iB & C+iD \\ -C+iD & A-iB \end{bmatrix}, A = -\cos(\psi_1)\cos(\psi_2),$$

$$B = -\sin(\psi_3)\sin(\psi_1), C = -\cos(\psi_1)\sin(\psi_2), D = -\sin(\psi_1)\cos(\psi_3),\qquad(1.19)$$

$$A^2+B^2+C^2+D^2 = 1, \psi_1 = \zeta - v - \xi/2, \psi_2 = \xi/2, \psi_3 = \xi/2 + v,$$

Here $v/2$, $\zeta/2$, and $(v+\xi)/2$ are the orientations of the first quarter-wave plate (QWP), half-wave plate and the second QWP at the vertical axis Y.

As follows from Equations (1.17) and (1.18):

$$T_1 = TB, T_1 = \begin{bmatrix} A_1+iB_1 & C_1+iD_1 \\ -C_1+iD_1 & A_1-iB_1 \end{bmatrix}, A_i^2+B_i^2+C_i^2+D_i^2 = 1.\qquad(1.20)$$

As follows from Equation (1.18), the condition of the SOP reproducibility in $n$-roundtrips takes the form:

$$\left(T_1 \exp(G)\right)^n = a\mathbf{I}, \; \mathbf{I} = \begin{bmatrix} 1 & 0 \\ 0 & 1 \end{bmatrix}, |a| = 1, \arg(a) = \pi k, k = 0,1. \qquad (1.21)$$

If we neglect SOP rotation caused by an active medium, i.e., $\exp(G) = I$, reproducibility of SOP for two roundtrips for the condition $A_1 = C_1 = D_1 = 0$ and $B_1 = 1$ means $L_b = 2L$ whereas for reproducibility for three round trips for condition $C_1 = D_1 = 0$ results in $A_1 = 3^{1/2}/2, B_1 = 1/2$, i.e. $L_b = 3L$.

To calculate pulse-to-pulse evolution of SOP numerically, we transform Equation (1.18) into the distributed form as follows:

$$\frac{d\Psi}{dt_s} = \left(G + \ln(T_1)\right)\Psi, \qquad (1.22)$$

Given the condition $C_1 = D_1 = 0$ for the case of SOP reproducibility in $n$-roundtrips, Equation (1.22) take the form

$$\frac{d\Psi}{dt_s} = G\Psi + \begin{pmatrix} i\pi L / L_{b1} & 0 \\ 0 & -i\pi L / L_{b1} \end{pmatrix}\Psi + NL, \qquad (1.23)$$

Where $L_{b1}$ is the beat length for combined fiber-POC birefringence, $NL$ describes the contribution of the Kerr nonlinearity as follows:

$$NL = i\frac{\gamma L I_{ss}}{2}\begin{pmatrix} |u|^2 u + \frac{2}{3}|v|^2 u + \frac{1}{3}v^2 u^* \\ |v|^2 v + \frac{2}{3}|u|^2 v + \frac{1}{3}u^2 v^* \end{pmatrix}. \qquad (1.24)$$

For an analytical study of SOP evolution, we substitute $\Psi = \left(|u|\exp(i\varphi_x), |v|\exp(i\varphi_y)\right)^T$ in Equation (1.23) and find the equation for the phase difference $\Delta\varphi = \varphi_x - \varphi_y$

$$\frac{d\Delta\varphi}{dt_s} = -\frac{2\pi L}{L_{b1}} + \frac{\gamma L I_{ss}}{12}\left(|v|^2 - |u|^2\right)\left(1 - 2\cos(2\Delta\varphi)\right) + Im(D_{yy}) - Im(D_{xx})$$
$$+ \frac{\left(|v|^2 - |u|^2\right)}{|u||v|}Im(D_{xy})\cos(\Delta\varphi) - \frac{\left(|v|^2 + |u|^2\right)}{|u||v|}Re(D_{xy})\sin(\Delta\varphi). \qquad (1.25)$$

## 1.10 SPIRAL POLARIZATION ATTRACTOR (THEORY)

Adjusting the in-cavity polarization controller enables changing fiber-POC birefringence from zero isotropic case to the case of high birefringent cavity. For the case of weak birefringence ($L_{b1} >> L$) and the pump power for laser below the threshold, cylindrical symmetry leads in SOP degeneration. However, for the pump power above the threshold value, instability of the steady state solution with the Stokes vector $\mathbf{S} = (S_0, 0, 0\pm1)$ results in the emergence of a double scroll attractor located at the Poincaré sphere (Sergeyev, 2014; Sergeyev et al., 2014).

Results for the numerical simulations of Equation (1.13) for the case of the isotropic cavity ($L_{b1} >> L$) are shown in Figure 1.24. Anisotropy in the cavity is caused by the elliptically polarized pump and the external fiber patchcord transforms the output lasing SOP. When the pump is circularly polarized ($\delta = 1$), the spiral attractor is symmetrical with the repeatable trajectories (Figure 1.24a). If the pump SOP is an elliptical ($\delta = 0.8$), the laser becomes more anisotropic and the output SOP is locked (Figure 1.24b). Increased pump power and weak deviation of the pump SOP from the circular ($\delta = 0.99$) transforms the symmetry and the trajectories fill more densely the surface of the Poincare sphere (Figure 1.24c).

Given the polarimeter's photodetector has a cut-off frequency of 1 MHz which means that, averaged over 25 roundtrips, the attractors in Figure 1.24 can have the different shape. To justify such a comment, we have processed the time domain waveforms shown in Figure 1.24 by using a low-pass filter in the form of a Hanning window (Transmission spectrum $T(f) = (1 + \cos(\pi f / f_c))/2$, $f \le f_c$) As a result, the spiral attractor in Figure 1.24a is slightly modified toward the shape similar to the experimentally observed (Figures 1.2a and 1.25a). Unlike this case, the filtering does not affect a polarization locked case (Figure 1.24b and 1.25b). In addition, the low-pass filter transforms the fast-evolving trajectory in Figure 1.24c to the slow-evolving one which is taking the shape of the double semi-circle trajectory similar to the experimentally observed (Figure 1.25c and 1.2c). Presented in Figures 1.24a and 1.25a spiral attractor dynamics is related to the relaxation oscillations with the period in terms of round-trip time and notations to Equation (1.12) of $T_{osc} \sim \varepsilon^{1/2}$ (Khanin, 2005). With 40 ns round trip time and $\varepsilon = 10^{-4}$, we have $T_{osc} \sim 4$ μs. Neglecting the gain dynamics leads to $\varepsilon \to \infty$ in Equation (1.12) and so $T_{osc} \to \infty$. It means that auto-oscillations doesn't exist and so there is no spiral attractor. Thus, Equation (1.12) can't be further simplified and slow gain dynamics has to be included into consideration.

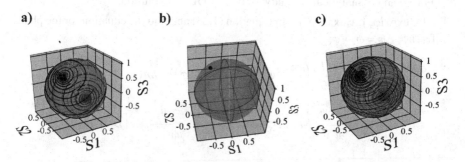

**FIGURE 1.24** Theoretically obtained: (a, c) Polarization precessing; and (b) Polarization locked vector solitons in terms of the normalized Stokes parameters at the Poincaré sphere. Parameters: a–c, $\varepsilon = 10^{-4}$, $\alpha_1 L = 200/\ln(10)$, $\alpha_2 L = 0.136$, $\alpha_3 = 10^{-4}$, $\alpha_4 L = 50/\ln(10)$, $\chi = 5/3$, $\Delta = 0.1$, $\gamma L I_{ss} = 2 \times 10^{-6}$; a, $I_p = 30$, $\delta = 1$; b, $I_p = 30$, $\delta = 0.8$; c, $I_p = 100$, $\delta = 0.99$. (Adapted from Sergeyev et al., 2014.)

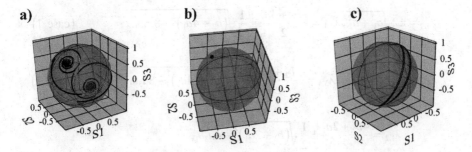

**FIGURE 1.25** Low-pass filtering with a Hanning window: (a) Spiral attractor after filtering; (b) Locked SOP; (c) Transformation of the double-scroll attractor to the double semi-circle by data filtering; (Transmission spectrum $T(f) = \left(1 + \cos\left(\pi f / f_c\right)\right)/2, \; f \le f_c = 1\text{MHz}$). Parameters: a–f, h–i, $\varepsilon = 10^{-4}$, $\alpha_1 L = 200/\ln(10)$, $\alpha_2 L = 0.136$, $\alpha_3 = 10^{-4}$, $\alpha_4 L = 50/\ln(10)$, $\chi = 5/3$, $\Delta = 0.1$, $\gamma L I_{ss} = 2 \times 10^{-6}$; a–c, $I_p = 30$, $\delta = 1$; d–f, $I_p = 100$, $\delta = 0.99$; h, i, $I_p = 30$, $\delta = 0.8$. (Adapted from Sergeyev et al., 2014.)

In the theory of coupled oscillators', the weak coupling leads to a complex behavior, while the strong coupling leads to the quenching oscillations and emerging the globally stable steady state (the Bar-Eli effect (Aronson et al., 1990)). For our vector model, the coupling strength depends on the pump SOP ellipticity and power. Also, coupling the orthogonally polarized lasing SOPs takes place through the gain sharing, detuning of the lasing wavelength with respect to the maximum of the gain spectrum and the Kerr nonlinearity.

To find the contribution of each of these factors to the origin of the spiral attractor, we present three different cases: (1) Scalar model of the Er-doped active medium, and a vector model of CNT; (2) A scalar model of both Erbium active medium and CNT; (3) A vector model of the active Erbium medium, and a scalar model of CNT at $\Delta = 0$ ($\Delta$ is detuning of the lasing wavelength with respect to the maximum of the gain spectrum).

First, we linearized the equations by substituting $|u| = |\mu_0| + x_1, |v| = |v_0| + x_1$ $\Delta\varphi = \Delta\varphi_0 + x_3, f_1 = f_{10} + x_4, \quad f_2 = f_{20} + x_5, \quad f_3 = f_{30} + x_5$ in Equation (1.12) and account for the different coefficients for the cases (1)–(3). Here $|u_0|, |v_0|, \Delta\varphi_0, f_{10}, f_{20}, f_{30}$ are steady state solutions for the cases (1)–(3)

$$|u_0|^2 = |v_0|^2 = \frac{\pi\alpha_1}{2\chi\alpha_4}\left(\frac{I_p}{2}(\chi-1)-1\right) - \frac{\pi\left(1+\Delta^2\right)}{2\chi}\left(1+\frac{I_p}{2}\right),$$

$$\Delta\varphi_0 = \pm\frac{\pi}{2}, f_{10} = \frac{I_p(\chi-1)/2-1}{1+\dfrac{I_p}{2}+\dfrac{2\chi|u_0|^2}{\left(1+\Delta^2\right)\pi}}, f_{20} = f_{30} = 0. \tag{1.26}$$

Where $u = |u|\exp\left(\varphi_u\right)$, $v = |v|\exp\left(\varphi_v\right)$, and $\Delta\varphi = \varphi_u - \varphi_v$ is the phase difference between two linearly polarized SOPs. Substituting $x_i = \tilde{x}_i\exp\left(\lambda t_s\right)$ into the linearized equations, we find the following eigenvalues

$$\lambda_{1,2} = \pm 2\sqrt{a_1^2 + a_2^2}, \lambda_{3,4} = \frac{-b_2 + 4a_1}{2} \pm \frac{1}{2}\sqrt{\left(b_2 + 4a_1\right)^2 - 8a_3b_1}. \qquad (\text{case } 1)$$

$$\lambda_1 = 2a_2, \lambda_2 = -2a_2, \lambda_{3,4} = \frac{-b_2 + 4a_1}{2} \pm \frac{1}{2}\sqrt{\left(b_2 + 4a_1\right)^2 - 8a_3b_1}. \qquad (\text{case } 2)$$

$$\lambda_{1,2} = \frac{-b_2 + 2a_1}{2} \pm \frac{1}{2}\sqrt{\left(b_2 + 2a_1\right)^2 - 4a_3b_1},$$

$$\lambda_{3,4} = \frac{-b_2 - 4a_1}{2} \pm \frac{1}{2}\sqrt{\left(b_2 - 4a_1\right)^2 - 8a_3b_1}, \qquad (\text{case } 3) \qquad (1.27)$$

$$\lambda_{5,6} = \frac{-b_2 + 4a_1}{2} \pm \frac{1}{2}\sqrt{\left(b_2 + 4a_1\right)^2 - 8a_3b_1}.$$

Where

$$a_1 = \frac{\alpha_2 L \alpha_3 |u_0|^2}{2\pi},$$

$$a_2 = \frac{2\gamma L I_{ss} |u_0|^2}{3}, a_3 = \frac{\alpha_1 L}{1+\Delta^2},$$

$$b_1 = \varepsilon \chi f_{10} \frac{2|u_0|^2}{\left(1+\Delta^2\right)\pi}, \qquad (1.28)$$

$$b_2 = \varepsilon \left( 1 + \frac{I_p}{2} + \frac{2\chi |u_0|^2}{\left(1+\Delta^2\right)\pi} \right)$$

Next, we introduce a saddle index as follows (Ovsyannikov and Shilnikov, 1987)

$$v = |\rho / \gamma|, \text{ If } \lambda_1 = \gamma > 0, \lambda_{2,3} = -\rho \pm i\omega, (\omega \neq 0, \rho > 0), \text{ or}$$
$$\lambda_1 = -\gamma < 0, \lambda_{2,3} = \rho \pm i\omega, (\omega \neq 0, \rho > 0). \qquad (1.29)$$

Equation (1.29) describe saddle-focus with limit cycles emerging in the homo-clinic bifurcation. According to the Shilnikov theorem, the stability and number of the limit cycles is defined by saddle index $v$ (Ovsyannikov and Shilnikov, 1987). If $v$ >1 homoclinic bifurcations results in one stable limit cycle. Unlike this, for $v$ <1 an infinite number of unstable cycles emerge and form a chaotic attractor (Ovsyannikov and Shilnikov, 1987). The saddle index as a function of pump power $I_p$ is shown in Figure 1.26. As follows from Figure 1.26, the double-scroll attractor cannot exist for the first and the second case and, also, for the third case where $\lambda_i = -\rho_i \pm i\omega_i, \quad (\omega_i \neq 0, \rho_i > 0)$. Thus, we conclude that the detuning of the lasing wavelength with respect to the maximum of the gain spectrum and SOPs coupling through the gain sharing leads to the complex dynamics.

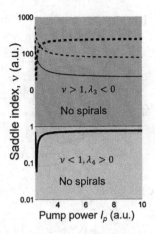

**FIGURE 1.26** Saddle index $\nu$ as a function of the normalized pump power $I_p$ for the first case (solid line) and the second case (dashed line). Parameters (thin and thick lines): $\varepsilon = 10^{-4}$, $\alpha_1 L = 200/\ln(10)$, $\alpha_2 L = 0.136$, $\alpha_4 L = 50/\ln(10)$, $\chi = 5/3$, $\Delta = 0.1$, $\delta = 1$, $\gamma L I_{ss} = 2 \times 10^{-6}$; (thin lines): $\alpha_3 = 10^{-4}$; (thick lines): $\alpha_3 = 10^{-2}$ (Adapted from Sergeyev et al., 2014).

## 1.11 INTERPLAY BETWEEN POLARIZATION HOLE BURNING AND IN-CAVITY BIREFRINGENCE (THEORY)

To illustrate the interplay between birefringence and polarization hole burning, we solve Equation (1.12) numerically by varying pump SOP ellipticity $\delta$ and beat length $L_{b1}$, and using parameters values quite close to the experimental ones (Sergeyev, 2014), $viz.$ $L = 10$ m, $\alpha_1 L = \ln(10)6.4$, $\alpha_2 L = 0.136$, $\alpha_3 = 10^{-4}$, $\alpha_4 L = \ln(10)0.5$, $\chi = 3/2$, $\Delta = 0.1$, $I_p = 30$, $\gamma L I_{ss} = 2 \times 10^{-6}$, $\varepsilon = 10^{-4}$. The results for $\delta = 1$ (circularly polarized pump) are shown in Figure 1.27.

As follows from Figure 1.27a–f, weak birefringence can distort spiral attractor and results in SOP localization close to the circle $s_2^2 + s_3^2 = 1$. In line with Equation (1.21), SOP is reproduced in $n$ round trips with a drift caused by polarization hole burning.

By changing ellipticity from to $\delta = 1$ to $\delta = 0.5$, we find that SOP is localized at the circle $s_2^2 + s_3^2 = a$ ($a < 1$, $s_1 \neq 0$) with slightly suppressed SOP drift (Figure 1.28a, b). For the pump power decreased from $I_p = 30$ to $I_p = 20$, the drift is completely suppressed (Figure 1.28c). The origin of drift can be clarified by analyzing the signal power $S_0$ as a function of number of round trips (Figure 1.28d–f). As follows from Figure 1.28d, e, fast oscillations of $S_0$ are caused by changes in the pulse-to-pulse rotation matrix associated with gain $G$ (Equation 1.23). Elliptically polarized pump with ellipticity of $\delta = 0.5$ leads to the light-induced anisotropy and so suppressed oscillations and (Figure 1.28e). As follows from Figure 1.28e. Stokes parameter $s_1$ increase with increased light-induced anisotropy that, according to Equation (1.25)

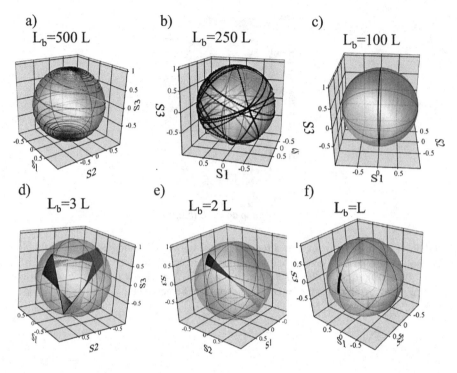

**FIGURE 1.27** Fast and slow evolution of vector solitons in terms of Stokes parameters in the Poincaré sphere. Parameters: $L = 10$ m, $\alpha_1 L = \ln(10)6.4$, $\alpha_2 L = 0.136$, $\alpha_3 = 10^{-4}$, $\alpha_4 L = \ln(10)0.5$, $\chi = 3/2$, $\Delta = 0.1$, $I_p = 30$, $\gamma L I_{ss} = 2 \times 10^{-6}$, $\varepsilon = 10^{-4}$, $\delta = 1$; (a) $L_b = 500$ L; (b) $L_b = 250$ L; (c) $L_b = 100$ L; (d) $L_b = 3$ L; (e) $L_b = 2$ L; (f) $L_b = L$. (Reprinted with permission from Sergeyev, 2014.)

results in decreased contribution of the active medium to the SOP drift. By decreasing the pump power from $I_p = 30$ to $I_p = 20$ we reduce the lasing powers $|v|^2$, $|u|^2$ that results in suppression of $S_0$ (normalized output power) oscillations and so, according to Equation (1.25), in suppression of SOP drift as shown in Figure 1.28c, f.

The obtained theoretical results are in a good agreement with our experimental data obtained previously for fundamental, bound state (BS) and multipulsing (MP) soliton operations (Sections 1.1–1.4 (Sergeyev et al., 2012; Mou et al., 2013; Tsatourian et al., 2013a, b; Sergeyev, 2014; Sergeyev et al., 2014)). For example, application of ansatz (Equation 1.11) is justified by our experimental study where pulse width is fixed (Sections 1.1–1.4). Also, the ansatz can be used for different pulse shapes with fixed pulse widths also, viz. for Gaussian in the case of normal dispersion, multipulsing and bound soliton regimes (Sergeyev et al., 2012; Mou et al., 2013; Tsatourian et al., 2013a, b; Sergeyev, 2014; Sergeyev et al., 2014)).

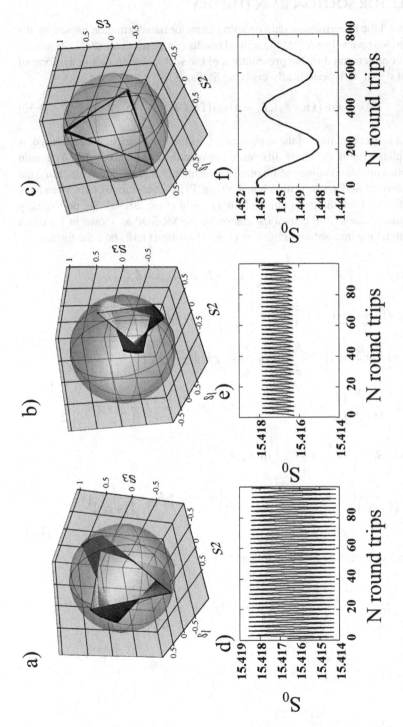

**FIGURE 1.28** Interplay between different types of birefringence caused by in-cavity polarisation controller and polarization hole burning, and light-induced anisotropy caused by elliptically polarized pump; (a)–(c) Fast and slow evolution of vector solitons in in terms of Stokes parameters in the Poincaré sphere; (d)–(e) output power signal $S_0$ as a function of number of round trips. Parameters: (a)–(f) $L_b = 3$ L; (a), (d) $I_p = 30$, $\delta = 1$; (b), (e) $I_p = 20$, $\delta = 0.5$; (c), (f) $I_p = 20$, $\delta = 1$. (Reprinted with permission from Sergeyev, 2014.)

## 1.12 VECTOR SOLITON RAIN (THEORY)

To understand the experimental data on spiral attractor transformation caused by the soliton rain, we review our recent theoretical results (Sergeyev et al., 2022). Equation (1.12) are complemented by the presentation of the vector soliton rain in the form of an injected signal with periodically evolving SOPs:

$$E_x = a \cdot \cos\left(\Omega t + \phi_0\right), E_y = a \cdot \sin\left(\Omega t + \phi_0\right) \cdot \exp\left(\Delta\varphi\right). \tag{1.30}$$

Where $a$ is the amplitude of the soliton rain, $\Omega$ is the frequency of oscillations, $\phi_0$ is the initial phase, $\Delta\varphi$ is the phase difference between the orthogonal SOPs. The main pulse depletes orientation distribution of inversion mainly at orientation coinciding the linearly polarized SOP (polarization hole burning, PHN) in and so SR pulses can have a SOPs different from the main pulse's SOP (Kbashi et al., 2019a). So, periodically evolving main pulse's SOP causes oscillation of the SR SOP as shown in Equation (1.30). With taking into account Equation (1.30), Equation (1.12) take the form:

$$\frac{du}{dt} = i\beta u + i\frac{\gamma}{2}\left(|u|^2 u + \frac{2}{3}|v|^2 u + \frac{1}{3}v^2 u^*\right) + D_{xx}u + D_{xy}v + E_x,$$

$$\frac{dv}{dt} = -i\beta v + i\frac{\gamma}{2}\left(|v|^2 v + \frac{2}{3}|u|^2 v + \frac{1}{3}u^2 v^*\right) + D_{xy}u + D_{yy}v + E_y,$$

$$\frac{dn_0}{dt} = \varepsilon\left[I_p + 2R_{10} - \left(1 + \frac{I_p}{2} + \chi R_{10}\right)n_0 - \chi R_{11}n_{12} - \chi n_{22}R_{12}\right],$$

$$\frac{dn_{12}}{dt} = \varepsilon\left[\frac{\left(1-\delta^2\right)}{\left(1+\delta^2\right)}\frac{I_p}{2} + R_{11} - \left(1 + \frac{I_p}{2} + \chi R_{10}\right)n_{12} - \left(\frac{\left(1-\delta^2\right)}{\left(1+\delta^2\right)}\frac{I_p}{2} + \chi R_{11}\right)\frac{n_0}{2}\right],$$

$$\frac{dn_{22}}{dt_s} = \varepsilon\left[R_{12} - \left(1 + \frac{I_p}{2} + \chi R_{10}\right)n_{22} - \chi R_{12}\frac{n_0}{2}\right],$$

$$R_{10} = \frac{1}{\left(1+\Delta^2\right)}\left(|u|^2 + |v|^2\right), R_{11} = \frac{1}{\left(1+\Delta^2\right)}\left(|u|^2 - |v|^2\right), R_{12} = \frac{1}{\left(1+\Delta^2\right)}\left(uv^* + vu^*\right),$$

$$\tag{1.31}$$

Coefficients $D_{ij}$ can be found as follows:

$$D_{xx} = \frac{\alpha_1\left(1-i\Delta\right)}{1+\Delta^2}\left(f_1 + f_2\right) - \alpha_2 + \ln\left(1 - \frac{\alpha_0}{1+\alpha_s\left(|u|^2 + v^2\right)}\right),$$

$$D_{yy} = \frac{\alpha_1\left(1-i\Delta\right)}{1+\Delta^2}\left(f_1 - f_2\right) - \alpha_2 + \ln\left(1 - \frac{\alpha_0}{1+\alpha_s\left(|u|^2 + v^2\right)}\right), \tag{1.32}$$

$$D_{xy} = D_{yx} = \frac{\alpha_1\left(1-i\Delta\right)}{1+\Delta^2}f_3.$$

To obtain results shown in Figure 1.29, we used the following parameters: (a), (d), (g) $I_p = 25$, $a = 0.001$; (b), (e), (h) $I_p = 22$, $a = 0.02$; (c), (f), (i) $I_p = 30$, $a = 10$. The other parameters: (a)–(i) $\beta_L = \beta_C = 0$, $\alpha_1 = 5.38$, $\alpha_2 = 10^{-3}$, $\alpha_0 = 0.136$, $\delta = 0.99$ (elliptically polarized pump SOP), $D = 0.13$ $\varepsilon = 10^{-4}$, $\chi_p = 1$, $\chi_s = 2.3$, $\Omega = 0.005\pi$, $\phi_0 = 0$, $\Psi = \pi/2$. To model the effect of the output SOP transformation caused by the patchcord connected to the polarimeter, we use the 3D rotation (around axes related to the Stokes parameters $S_0$, $S_1$, $S_2$, $S_3$) matrix (Varshalovich et al., 1988):

$$
\begin{pmatrix} \tilde{S}_1 \\ \tilde{S}_2 \\ \tilde{S}_3 \\ \tilde{S}_4 \end{pmatrix} = \begin{bmatrix} a_{11} & a_{12} & a_{13} & 0 \\ a_{21} & a_{22} & a_{23} & 0 \\ a_{31} & a_{32} & a_{33} & 0 \\ 0 & 0 & 0 & 1 \end{bmatrix} \begin{pmatrix} S_1 \\ S_2 \\ S_3 \\ S_0 \end{pmatrix},
$$

$$
\begin{aligned}
a_{11} &= \cos(\psi)\cos(\gamma), \quad a_{12} = \cos(\gamma)\sin(\alpha)\sin(\psi) - \cos(\alpha)\sin(\gamma), \\
a_{13} &= \cos(\alpha)\cos(\gamma)\sin(\psi) + \sin(\alpha)\sin(\gamma), \\
a_{21} &= \cos(\psi)\sin(\gamma), a_{22} = \cos(\alpha)\cos(\gamma) + \sin(\alpha)\sin(\psi)\sin(\gamma), \\
a_{23} &= -\cos(\gamma)\sin(\alpha) + \sin(\psi)\sin(\gamma), \\
a_{23} &= -\sin(\gamma), \quad a_{23} = \cos(\psi)\sin(\alpha) \\
a_{33} &= \cos(\alpha)\cos(\psi).
\end{aligned}
\tag{1.33}
$$

As follows from Figure 1.29a–h, the injected signal with evolving SOP modifies the spiral attractor and polarization dynamics of the SOPs x and y. Theoretical results are quite close to the experimental data (Figure 1.15a–h) for the soliton rain's condensed phase in Figure 1.14a, b. Increased amplitude of the injected signal from $a = 0.02$ to $a = 10$ transforms the spiral attractor to the circle (Figure 1.29c), and x and y SOPs' oscillations take the form of antiphase close to harmonic oscillations with the fast phase difference switching (Figure 1.29f, i). The dynamics is quite close to the experimental results shown in Figure 1.15c, f, i and the corresponding the soliton rain bunch shown in Figure 1.14c.

Kbashi and co-workers demonstrated that the mechanism driving the SR origin and merging to the condensate phase is competition between polarization hole burning (PHB) caused by SR pulses and holes refilling by the pump wave and active medium (Kbashi et al., 2019a). Here we reveal a new effect of the soliton rain on the active medium in the context of modifying the polarization properties. In the SR condensed phase (Figure 1.14a, b, d, e), the SR completely depletes population inversion and cw component can not appear. The active medium is slightly modified by SHB through the induction of a small circular birefringence and so the spiral attractor on the Poincaré sphere emerges (Figure 1.14a, b, d, e, g, h and Figure 1.29b, e, h). The soliton bunch appears when PHB cannot deplete the population inversion completely, and CW components generated (Figure 1.14f). The soliton bunch has a large amplitude and the periodically evolving SOP caused by the its drift and so transforms the spiral attractor to the circle on the Poincaré sphere (Figures 1.15c, f, i and 1.29c, f, i). The obtained results can pave the wave to development a new technique for the

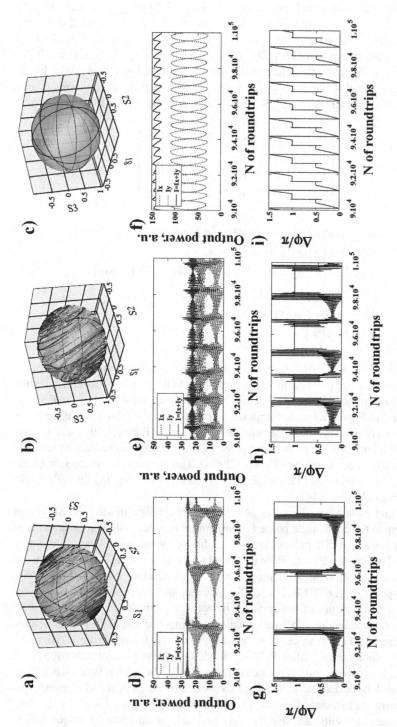

**FIGURE 1.29** Slow polarization dynamics: (a)–(c) trajectories on the Poincaré sphere.); (d)–(f) The output power vs number of the round trips for two linearly cross-polarized SOPs $I_x$ (dashed line) and $I_y$ (dotted line) and total power $I = I_x + I_y$ (black).); (g–f) the phase difference vs number of the round trips. Parameters: $\alpha = -\pi / 4, \beta = \pi / 4, \gamma = 2\pi / 5$, (a–i); $I_p = 25$ (a, d, g); $I_p = 22$ (b, e, h); $I_p = 30$ (c, f, i). The time span of $10^4$ corresponds to 1 ms. (Reprinted with permission from Sergeyev et al., 2022.)

laser dynamics' control by using the injected optical signal with evolving states of polarization that of interest for different application including spectroscopy, metrology and biomedical diagnostics.

## 1.13 VECTOR BRIGHT-DARK ROGUE WAVES (THEORY)

To justify mechanism of the RWs emergence at the fast and slow time scales, we modify Equation (1.12) as follows:

$$\frac{dS_0}{dt} = \left( \frac{2\alpha_1 f_1}{1+\Delta^2} - 2\alpha_2 - \ln\left( 1 - \frac{\alpha_0}{1+\alpha_s S_0} \right) \right) S_0 + \frac{2\alpha_1 f_2}{1+\Delta^2} S_1 + \frac{2\alpha_1 f_3}{1+\Delta^2} S_2,$$

$$\frac{dS_1}{dt} = \gamma S_2 S_3 + \frac{2\alpha_1 f_2}{1+\Delta^2} S_0 + \left( \frac{2\alpha_1 f_1}{1+\Delta^2} - 2\alpha_2 - \ln\left( 1 - \frac{\alpha_0}{1+\alpha_s S_0} \right) \right) S_1 - \beta_c S_2 - \left( \frac{2\alpha_1 f_3 \Delta}{1+\Delta^2} \right) S_3.$$

$$\frac{dS_2}{dt} = -\gamma S_1 S_3 + \frac{2\alpha_1 f_3}{1+\Delta^2} S_0 + \beta_c S_1 + \left( \frac{2\alpha_1 f_1}{1+\Delta^2} - 2\alpha_2 - \ln\left( 1 - \frac{\alpha_0}{1+\alpha_s S_0} \right) \right)$$

$$\times S_2 + \left( \frac{2\alpha_1 f_2 \Delta}{1+\Delta^2} \right) S_3,$$

$$\frac{dS_3}{dt} = \left( \frac{2\alpha_1 \Delta f_3}{1+\Delta^2} \right) S_1 - \left( \frac{2\alpha_1 \Delta f_2}{1+\Delta^2} \right) S_2 + \left( \frac{2\alpha_1 f_1}{1+\Delta^2} - 2\alpha_2 - \ln\left( 1 - \frac{\alpha_0}{1+\alpha_s S_0} \right) \right) S_3,$$

$$\frac{df_1}{dt} = \varepsilon \left[ \frac{(\chi_s - 1) I_p}{2} - 1 - \left( 1 + \frac{I_p \chi_p}{2} + d_1 S_0 \right) f_1 - \left( \frac{I_p \chi_p}{2} \frac{(1-\delta^2)}{(1+\delta^2)} \right) f_2 - d_1 S_2 \times f_3 \right],$$

$$\frac{df_2}{dt} = \varepsilon \left[ \frac{(1-\delta^2)}{(1+\delta^2)} \frac{I_p (\chi_s - 1)}{4} - \left( \frac{I_p \chi_p}{2} + 1 + d_1 S_0 \right) f_2 - \left( \frac{(1-\delta^2)}{(1+\delta^2)} \frac{I_p \chi_p}{2} + d_1 S_1 \right) \frac{f_1}{2} \right],$$

$$\frac{df_3}{dt} = -\varepsilon \left[ \frac{d_1 S_2 f_1}{2} + \left( \frac{I_p \chi_p}{2} + 1 + d_1 S_0 \right) \times f_3 \right].$$

$$(1.34)$$

Here time and length are normalized to the roundtrip and cavity length, respectively; $S_i$ ($i = 0,1,2,3$) are the Stokes parameters defined in Equation (1.1) ($S_0$ is the output power, pump and lasing powers are normalized to the corresponding saturation powers); $\beta_{L(C)} = 2\pi / L_{bL(bC)}$ is the linear (circular) birefringence, $L_{bL(bC)}$ is the linear (circular) birefringence beat length; $\alpha_1$ is the total absorption of erbium ions at the lasing wavelength, $\alpha_2$ is the total insertion losses in the cavity; $\delta$ is the ellipticity of the pump wave, $\varepsilon = \tau_R/\tau_{Er}$ is the ratio of the round trip time $\tau_R$ to the lifetime of erbium ions at the first excited level $\tau_{Er}$; $\chi_{p,s} = (\sigma_a^{(s,p)} + \sigma_e^{(s,p)})/\sigma_a^{(s,p)}, (\sigma_a^{(s,p)}$ and $\sigma_e^{(s,p)}$ are absorption and emission cross-sections at the lasing (s) and pump (p) wavelengths); $\Delta$ is the detuning of the lasing wavelength with respect to the maximum of the gain spectrum (normalized to the gain spectral width); $d_1 = \chi_s / \pi (1+\Delta^2)$.

To explain mechanism of RW emergence, we solve Equation (1.34) numerically by using the parameters shown in Figure 1.30.

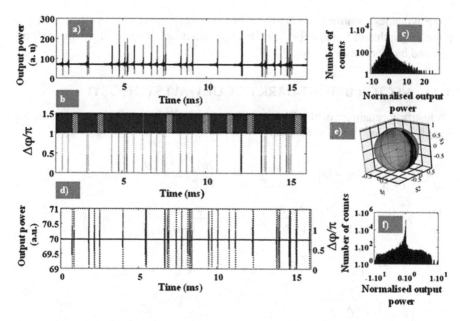

**FIGURE 1.30** Laser dynamics averaged over the roundtrip time and after low pass filtering. Dynamics before (a–c) and after (d–f) low-pass filtering with a Hanning window with transmission spectrum ($T(f)=(1+\cos(\pi f/f_c))/2$, $f \leq f_c$ = 1 MHz ). (a) Dynamics of the output power; (b) Dynamics of the phase difference $\Delta\phi$; (c) Probability distribution histograms (the output power I is normalized as shown in Figure 1.2); (d) Dynamics of the output powers $I=I_x+I_y$ (black) and the phase difference $\Delta\phi$ (gray); (e) Trajectories on the Poincaré sphere; (f) Probability distribution histogram for the total output power (output power is normalized as shown in Figures 1.16 and 1.17). Parameters: $I_p$ = 76; $\alpha_1$ = 21.53, $\alpha_0$ = 0.136, $\alpha_s$ = 1.8 × 10$^{-5}$, $\alpha_2$ = 2.533, $\chi$ = 2.3, $\Delta$ = 0.1, $\gamma$ = 2 × 10$^{-6}$, $\varepsilon$ = 10$^{-4}$; (a–f) $\delta$ = 0.84 (elliptically polarized pump), $\beta$ = 1.1$\pi$ × 10$^{-2}$ (Reprinted with permission from Kbashi et al., 2018.)

As follows from Figure 1.30, adjusting the pump power, birefringence and the ellipticity of the pump wave leads to the emergence of the bright-dark rogue waves. After the data averaging over 30 roundtrips (1 MHz low-pass filter) the data are transformed (Figure 1.30d–f). The obtained results in Figure 1.30 are in a good correspondence to the experimental data shown in Figures 1.16 and 1.17. The anomalous spikes-dips in the output power (Figure 1.30a, b, d) coincides with the phase difference jumps in $\pi$ (Figure 1.30b, d), i.e., transitions between orthogonally polarized SOPs (Figure 1.30e). Probability distribution diagram for the total power $I = I_x + I_y$ (Figure 1.30c, f) is showing the presence the dark-bright RWs. The emergence of the RWs at the slow time scale is explained in terms of transition of the synchronization scenario from the phase locking to the chaotic phase drift (Section 1.16).

Thus, we demonstrate a new type of the bright-dark rogue waves resulting from the interaction between the orthogonal SOPs in an Er-doped mode-locked fiber laser. By adjusting in-cavity and the pump wave polarization controllers, we enable control of the coupling between SOPs towards SOPs' desynchronization and so emergence of the bright-dark rogue. The revealed mechanism shows a great potential for mapping conditions for RWs existence and so for developing techniques for suppression RWS in different distributed systems.

## 1.14 VECTOR RESONANCE MULTIMODE INSTABILITY (THEORY)

The experimental results on mode-locking based on Vector Resonance Multimode Instability (VRMI) can be well understood based on the vector model of an Er-doped fiber laser derived by (Sergeyev et al., 2018):

$$
\frac{\partial S_0}{\partial z} + \frac{\partial S_0}{\partial t} = \left( \frac{2\alpha_1 f_1}{1+\Delta^2} - 2\alpha_2 \right) S_0 + \frac{2\alpha_1 f_2}{1+\Delta^2} S_1 + \frac{2\alpha_1 f_3}{1+\Delta^2} S_2,
$$

$$
\frac{\partial S_1}{\partial z} + \frac{\partial S_1}{\partial t} = \gamma S_2 S_3 + \left( \frac{2\alpha_1 f_1}{1+\Delta^2} - 2\alpha_2 \right) S_1 + \frac{2\alpha_1 f_2}{1+\Delta^2} S_0 - \frac{2\alpha_1 f_3 \Delta}{1+\Delta^2} S_3,
$$

$$
\frac{\partial S_2}{\partial z} + \frac{\partial S_2}{\partial t} = -\gamma S_1 S_3 + \frac{2\alpha_1 f_3}{1+\Delta^2} S_0 + \left( \frac{2\alpha_1 f_1}{1+\Delta^2} - 2\alpha_2 \right) S_2 + \left( \frac{2\alpha_1 f_2 \Delta}{1+\Delta^2} - 2\beta \right) S_3,
$$

$$
\frac{\partial S_3}{\partial z} + \frac{\partial S_3}{\partial t} = \frac{2\alpha_1 \Delta f_3}{1+\Delta^2} S_1 - \frac{2\alpha_1 \Delta f_2}{1+\Delta^2} S_2 + 2\beta S_2 + \left( \frac{2\alpha_1 f_1}{1+\Delta^2} - 2\alpha_2 \right) S_3,
$$

$$
\frac{df_1}{dt} = \varepsilon \left[ \frac{(\chi_s - 1) I_p}{2} - 1 - \left( 1 + \frac{I_p \chi_p}{2} + d_1 S_0 \right) f_1 - \left( d_1 S_1 + \frac{I_p}{2} \xi \right) f_2 - d_1 S_2 f_3 \right],
$$

$$
\frac{df_2}{dt} = \varepsilon \left[ \xi \frac{I_p (\chi_s - 1)}{4} - \left( \frac{I_p \chi_p}{2} + 1 + d_1 S_0 \right) f_2 - \left( \frac{I_p \chi_p}{2} \xi + d_1 S_1 \right) \frac{f_1}{2} \right],
$$

$$
\frac{df_3}{dt} = -\varepsilon \left[ \frac{d_1 S_2 f_1}{2} + \left( \frac{I_p \chi_p}{2} + 1 + d_1 S_0 \right) f_3 \right].
$$

$$(1.35)$$

Here $\xi = (1-\delta^2)/(1+\delta^2)$ is parameter of the pump anisotropy.

Given the experimental condition that used pump power of 18 mW is much less than values (approx. 800 mW) required for mode locking based on nonlinear polarization rotation (NPR), the presented model excludes NPR-based mode-locking (Lecaplain et al., 2014) . The pulse width in our experiments (Figure 1.19) was estimated to be of 40 ns and so we can use approximation where the second order dispersion can be neglected in the model of VMRI-based mode-locking. Also, the pulse width is much longer than the transverse relaxation time of 160 fs. Therefore, the medium polarization dynamics can be ignored (Fontana et al., 1995; Pessina et al., 1997; Pessina et al., 1999; Voigt et al., 2004; Lugiato et al., 2015).

To find conditions for the VRMI-based mode locking, we linearize the Equation (1.35) nearby the steady state solution $F_0 = \left( S_{00} \pm S_{00} 0\, 0\, f_{10} f_{20} 0 \right)^T$ and substitute the following ansatz into Equation (1.35):

$$
F(t,z) \equiv \left[ S_0 S_1 S_2 S_3 f_1 f_2 f_3 \right]^T = F_0 + \left[ x_0 x_1 x_2 x_3 x_4 x_5 x_6 x_7 \right]^T \exp(\lambda t + qz), \quad (1.36)
$$

where $F_0 = (S_{00} \pm S_{00}\ 0\ 0\ f_{10}\ f_{20}\ 0)^T$. As a result, we find the following equation for eigenvalues:

$$
\det
\begin{bmatrix}
a_i - iq - \lambda & 0 & 0 & 0 & a_2 & a_2 & 0 \\
-a_1 & a_1 - iq - \lambda & 0 & 0 & a_2 & a_2 & 0 \\
0 & 0 & a_1 - iq - \lambda & a_3 + a_4 & 0 & 0 & a_2 \\
0 & 0 & -a_3 & a_1 - iq - \lambda & 0 & 0 & \Delta \cdot a_2 \\
b_2 & b_4 & 0 & 0 & b_1 - \lambda & b_3 & 0 \\
b_4 & b_2/2 & 0 & 0 & b_3/2 & b_1 - \lambda & 0 \\
0 & 0 & b_2/2 & 0 & 0 & 0 & b_1 - \lambda
\end{bmatrix}
= 0
$$

$$(1.37)$$

where

$$
a_1 = \frac{2a_1}{1+\Delta^2} f_{10} - \alpha_2, a_2 = \frac{2\alpha_1}{1+\Delta^2} S_{00}, a_3 = \frac{2\alpha_1 \Delta}{1+\Delta^2} f_{20} - 2\beta, a_4 = -\gamma S_{00},
$$

$$
b_1 = -\varepsilon\left(1 + \frac{I_p \chi_p}{2} + d_1 S_{00}\right), b_2 = -\varepsilon d_1 f_{10}, b_3 = -\varepsilon\left(d_1 S_0 + \frac{I_p \chi_p \xi}{2}\right), b_4 = -\varepsilon d_1 f_{20},
$$

$$(1.38)$$

$$
S_{00} = \frac{-Q_2 - \sqrt{Q_2^2 - 4Q_1 Q_3}}{2Q_1}, Q_1 = -\frac{\alpha_2 d_1^2}{2},
$$

$$
Q_2 = -\alpha_2\left(2d_1\left(1 + I_p \chi_p\right) - \frac{d_1 I_p \xi}{2}\right) + \left(\frac{\chi_s - 1}{2} I_p - 1\right).
$$

As a result, Equation (1.37) has three branches of eigenvalues:

$$
\lambda - 2a_1 + 2iq = 0, \quad (I)
$$

$$
\lambda^3 + \left[-2b_1 + iq\right]\lambda^2 + \left(b_1^2 - \frac{b_3^2}{2} - \frac{3a_2 b_2}{2} - 2a_2 b_4 - 2iqb_1\right)\lambda + \frac{3a_2 b_1 b_2}{2} + 2a_2 b_1 b_4 -
$$

$$
2a_2 b_1 b_4 - a_2 b_2 b_3 - \frac{3a_2 b_3 b_4}{2} - \frac{ib_3^3 q}{2} + ib_1^2 q = 0, \quad (II)
$$

$$
\lambda^3 + \left(-2a_1 - b_1 + 2iq\right)\lambda^2 + \left(a_1^2 + 2b_1 a_1 + a_3^2 + a_4 a_3 - q^2 - \frac{a_2 b_2}{2} - 2ib_1 q - 2ia_1 q\right) +
$$

$$
b_1 q^2 - a_3^2 b_1 - a_1^2 b_1 + \frac{a_1 a_2 b_2}{2} - a_3 a_4 b_1 - \frac{a_2 a_3 b_2 \Delta}{2} - \frac{a_2 a_4 b_2 \Delta}{2} - \frac{ia_2 b_2 q}{2} + 2ia_1 b_1 q
$$

$$
= 0, \quad (III)
$$

$$(1.39)$$

As a result:

$$
\begin{aligned}
(I)\,\lambda_0 &= iq + A_0\left(I_p,\xi\right),\\
(II)\,\lambda_1 &= A_1\left(q,I_p,\xi\right)+i\Omega_1\left(q,I_p,\xi\right),\\
\lambda_2 &= A_2\left(q,I_p,\xi\right)+i\Omega_2\left(q,I_p,\xi\right),\\
\lambda_3 &= A_3\left(q,I_p,\xi\right)+i\Omega_3\left(q,I_p,\xi\right),\\
(III)\,\lambda_4 &= A_4\left(q,I_p,\beta,\xi\right)+i+\Omega_4\left(q,I_p,\beta,\zeta\right),\\
\lambda_5 &= A_5\left(q,I_p,\beta,\xi\right)+i\left(q+\Delta\Omega\left(q,I_p,\beta,\xi\right)\right),\\
\lambda_6 &= A_6\left(q,I_p,\beta,\xi\right)+i\left(q-\Delta\Omega\left(q,I_p,\beta,\xi\right)\right),\\
A_0\left(I_p,\xi\right)&>0, A_1\left(q,I_p,\xi\right)<0,\\
A_2\left(q,I_p,\xi\right)&<0,\\
A_3\left(q,I_p,\xi\right)&<0,\\
A_4\left(q,I_p,\xi\right)&<0,\\
A_5\left(q,I_p,\xi\right)&>0,\\
A_6\left(q,I_p,\xi\right)&>0.
\end{aligned}
\tag{1.40}
$$

Here $q = 0, \pm1, \pm2,\ldots\pm N$ is the wave number of the longitudinal mode and eigenvalues are normalized to the fundamental frequency $\omega = 2\pi/\tau_R$. The results are shown in Figure 1.31. The first and the second (multimode instability) thresholds are found in Figure 1.31a. For RNGH instability, the threshold pump powers for excitation of different number of longitudinal modes are different (Fontana et al., 1995; Pessina et al., 1997; Pessina et al., 1999; Voigt et al., 2004; Lugiato et al., 2015), whereas for the case of VRMI the powers are the same (Figure 1.31a). Also, the second threshold of the VRMI coincides with the first lasing threshold for circuitry polarized pump and slightly exceeds the first one with increased pump anisotropy parameter $\xi$. Threshold for MMI (branch I) coincides with the threshold for excitation of birefringence-dependent RF satellite lines (branch III). As follows from Figure 1.31b, the branch II corresponds to additional birefringence-independent spectral lines with the frequency splitting of 0.01f (f is the fundamental frequency) with respect to the longitudinal mode frequency q. The real part of the corresponding eigenvalues, i.e., $A_1$, $A_2$ and $A_3$, are less than zero, but the parametric phase locking with the frequencies of the branch I can activate the satellites. For branch III (Figure 1.31c), increased birefringence strength can result in the resonance conditions, i.e., matching longitudinal mode q satellites' frequencies the frequency of the longitudinal mode q+N (where N is integer) from the branch I.

As follows from Figures 1.31 and 1.19, the theoretical results are in a good correspondence with the experimental data. First, the threshold of the multimode instability (the second threshold) slightly exceeds the first lasing threshold (Figure 1.31a). Second, when the birefringence-dependence satellites frequencies deviates from the longitudinal mode frequencies (Figure 1.19a1–c1), the dynamics takes the form of complex oscillations in view of equal threshold conditions for all longitudinal modes

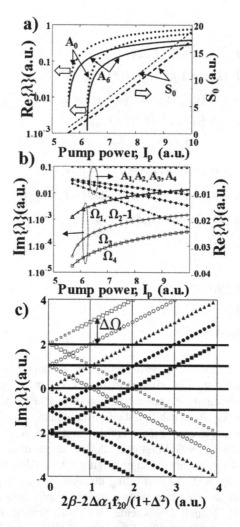

**FIGURE 1.31** The results of linear stability analysis of Equation (1.35). (a) Vector multi-mode instability in terms of positive real parts of eigenvalues, i.e., $A_0$ (dots), $A_5$ (solid line) ($A_6$ is close to the $A_5$ and so is not shown here) and the output signal $S_0$ (dashed line) vs pump power $I_p$ for $\xi = 0$ (upper lines), $\xi = 0.1$ (lower line); (b) The frequencies of the scalar branch: $\Omega_1$ (empty circles) $\Omega_2 - 1$ (empty triangles), $\Omega_3$ (empty squares) and vector branch $\Omega_4$ (empty diamonds) along with the real parts of the scalar branch: $A_1$ (filled circles), $A_2$ (filled triangles), $A_3$ (filled squares), and the vector branch: $A_4$ (filled diamonds) vs pump power $I_p$ for $\xi = 0.1$; (c) Frequencies for the branch I, i.e., $\Omega_0 = \pm q$, $q = 0,1,2,3$ (solid lines), and the vector branch (III), i.e., $\Omega_{5,6} = q \pm \Delta\Omega$ (squares, triangles, circles), vs the birefringence strength $2\beta - 2\Delta\alpha_1/(1+\Delta^2)$ for $\xi = 0.1$. The equality $\Omega_0 = \Omega_{5,6}$ for $2\beta - 2\Delta\alpha_1/(1+\Delta^2) = 1,2,\ldots$ means the resonance mode locking shown in Figure 1.31c. Parameters: $L = 615$ m, $2\alpha_1 = \ln(10)6.4$, $2\alpha_2 = \ln(10)0.5$, $\chi_s = 3/2$, $\chi_p = 1/0.7$, $\Delta = 0.1$, $I_p = 10$ (c), $\gamma = 2 \times 10^{-6}$, $\varepsilon = 10^{-3}$, $\beta = 1$ (a,b), q = 1 (b), $\xi = 0.1$ (b,c) (Reprinted with permission from Sergeyev et al., 2017).

and the absence of the synchronization (Figure 1.31a)). Finally, matching the frequencies for longitudinal modes' harmonics and the birefringence-dependent satellites leads their synchronization similar to the injection locking-based mode locking (Cundiff and Ye, 2003). Also, stable mode-locking is accompanied with the stable SOP locking Figures 1.18b, c. As follows from Figure 1.31, linear stability analysis results in the eigenfrequencies ratio with respect to the fundamental frequency as $1:10^{-1}:10^{-2}:10^{-3}:10^{-4}$. In addition, the presence of harmonics (more than 1,000, as follows from the experiment) requires accounting for many time scales and so the direct numerical simulation is very complex problem.

The demonstrated vector resonance multimode instability-based mode-locking can be potentially observed in the other distributed systems. The resonance of the satellite lines caused by the birefringence tuning with the other branch of eigenfrequencies leads to the synchronization phenomena which can be of interest in photonics and beyond.

## 1.15 VECTOR HARMONIC MODE-LOCKING (THEORY)

To understand the mechanism of harmonic mode-locking tunability and linewidth narrowing, we review our recently developed vector model of EDFLs (Sergeyev et al., 2021). The model accounts for the linear and circular birefringence and fast- and slow-axis modulation caused by $TR_{2m}$ acoustic modes. Without accounting for the gain dynamics, the SOP evolution in terms of the Stokes vector $S$ and number of roundtrips caused by the interplay of the factors mentioned above can be described as follows:

$$dS / dt = R \cdot W \times S, \tag{1.41}$$

Here time is normalized to the roundtrip time, $W = \left( \beta_L, 0, \beta_c \right)^T$ is the birefringence vector, $\beta_{L(C)} = 2\pi / L_{bL(bC)}$ is the linear (circular) birefringence strength, $L_{bL(bC)}$ is the beat length for linear (circular) birefringence. The matrix $R$ is the 3×3 matrix that defines the rotation of the birefringence vector around axis $OS_3$ caused by $TR_{2m}$ excitation (Collett, 2003):

$$R = \begin{bmatrix} \cos\left(\zeta\left(t\right)\right) & -\sin\left(\zeta\left(t\right)\right) & 0 \\ \sin\left(\zeta\left(t\right)\right) & \cos\left(\zeta\left(t\right)\right) & 0 \\ 0 & 0 & 1 \end{bmatrix}, \tag{1.42}$$

where $\zeta\left(t\right) = A_0 \cos\left(2\pi\Omega t\right)$. Here $\zeta\left(t\right)$ is the angle of the birefringence vector rotation, $A_0$ is the amplitude of rotation, and $\Omega$ is the frequency of oscillations at the $TR_{2m}$ acoustic mode. In Equations (1.41 and 1.42), the contribution of $TR_{2m}$ was accounting for only in the birefringence modulation context. The modulation of the refractive index was neglected. With accounting for Equations (1.41 and 1.42), along with the ASE noise in the cavity, Equations (1.30 and 1.31) take the following form (Sergeyev et al., 2021):

$$\frac{dS_0}{dt} = \left(\frac{2\alpha_1 f_1}{1+\Delta^2} - 2\alpha_2\right)S_0 + \frac{2\alpha_1 f_2}{1+\Delta^2}S_1 + \frac{2\alpha_1 f_3}{1+\Delta^2}S_2,$$

$$\frac{dS_1}{dt} = \gamma S_2 S_3 + \frac{2\alpha_1 f_2}{1+\Delta^2}S_0 + \left(\frac{2\alpha_1 f_1}{1+\Delta^2} - 2\alpha_2\right)S_1 - \beta_c S_2 - \left(\frac{2\alpha_1 f_3 \Delta}{1+\Delta^2} - \beta_L \sin\left(\zeta\left(t\right)\right)\right)S_3 + \sigma_1,$$

$$\frac{dS_2}{dt} = -\gamma S_1 S_3 + \frac{2\alpha_1 f_3}{1+\Delta^2}S_0 + \beta_c S_1 + \left(\frac{2\alpha_1 f_1}{1+\Delta^2} - 2\alpha_2\right)S_2 + \left(\frac{2\alpha_1 f_2 \Delta}{1+\Delta^2} - \beta_L \cos\left(\zeta\left(t\right)\right)\right)S_3 + \sigma_2,$$

$$\frac{dS_3}{dt} = \left(\frac{2\alpha_1 \Delta f_3}{1+\Delta^2} - \beta_L \sin\left(\zeta\left(t\right)\right)\right)S_1 - \left(\frac{2\alpha_1 \Delta f_2}{1+\Delta^2} - \beta_L \cos\left(\zeta\left(t\right)\right)\right)S_2 + \left(\frac{2\alpha_1 f_1}{1+\Delta^2} - 2\alpha_2\right)S_3 + \sigma_3,$$

$$\frac{df_1}{dt} = \varepsilon\left[\frac{(\chi_s - 1)I_p}{2} - 1 - \left(1 + \frac{I_p \chi_p}{2} + d_1 S_0\right)f_1 - \left(d_1 S_1 + \frac{I_p \chi_p}{2}\frac{(1-\delta^2)}{(1+\delta^2)}\right)f_2 - d_1 S_2 f_3\right],$$

$$\frac{df_2}{dt} = \varepsilon\left[\frac{(1-\delta^2)}{(1+\delta^2)}\frac{I_p(\chi_s - 1)}{4} - \left(\frac{I_p \chi_p}{2} + 1 + d_1 S_0\right)f_2 - \left(\frac{(1-\delta^2)}{(1+\delta^2)}\frac{I_p \chi_p}{2} + d_1 S_1\right)\frac{f_1}{2}\right],$$

$$\frac{df_3}{dt} = -\varepsilon\left[\frac{d_1 S_2 f_1}{2} + \left(\frac{I_p \chi_p}{2} + 1 + d_1 S_0\right) \times f_3\right].$$

(1.43)

$\sigma_i$ are Stokes parameters of the injected δ-correlated stochastic signal:

$$\langle\sigma_i(t)\rangle_t = 0, \quad \langle\sigma_i(t)\sigma_j(t-\tau)\rangle_t = \Sigma^2 \delta_{i,j}\delta(\tau), \quad \delta_{i,j} = \begin{cases} 1, i = j, \\ 0 \, i \neq j. \end{cases} \delta(\tau) = \begin{cases} \infty, \tau = 0, \\ 0 \, \tau \neq 0. \end{cases}$$

(1.44)

Here $\delta_{i,j}$ is the Kronecker symbol, $\delta(\tau)$ is the Dirac function, $\sigma^2 = 1/\pi_c$, $\tau_c$ is the correlation time.

To obtain results shown in Figure 1.32, we used the following parameters: (a)–(i) $\Omega = 7, A_0 = 0.1$; (a)–(c) $\beta_L = 2\pi/\sqrt{5}, \beta_C = 0$; (d)–(f) $\beta_L = 2\pi/\sqrt{5}, \beta_C = \pi\sqrt{2}/\sqrt{5}$; (g)–(i) $\beta_L = 2\pi/\sqrt{5}, \beta_C = \pi\sqrt{4}/\sqrt{5}$. The other parameters: (a)–(i) $\alpha_1 = 21.5, \alpha_2 = 2.53, I_p = 30, d = 0.5$ (elliptically polarized pump SOP), $\Delta = 0.1, \varepsilon = 10^{-4}, \chi_p = 1/0.75, \chi_s = 2.3, S = 10^{-3}$.

It was shown in Section 1.12 that tuning the linear birefringence results in the vector resonance multimode instability (Sergeyev et al., 2017). By adjusting the in-cavity and the pump wave polarization controllers, the birefringence strength can be increased and two satellite lines around the q-harmonic frequency emerge. Finally, longitudinal modes synchronization happens when the beat length equals the cavity length and so the frequencies of the satellites for q-harmonic are in resonance with the q+1-and q−1-harmonic.

The complexity of the vector model exceeds the complexity of any known scalar or vector models of fiber lasers considered elsewhere. Given the complexity of the problem, we use a few approximations to reveal the effect of the $TR_{2m}$ on the modulation of the output power at frequency $\Omega$. First, in the theoretical analysis, we accounted for the only interplay of the linear and circular birefringence with the $TR_{2m}$ acoustic mode-based modulation for harmonic $q = 0$ in terms of the ability of excitation of the output power oscillations at a frequency of $TR_{2m}$ mode.

The results of the theoretical analysis are shown in Figure 1.32a–i. As follows from Figure 1.32a, b, d, e the output power $I$ and $I_x$, $I_y$ are oscillating at frequency $\omega\sqrt{\beta_L^2 + \beta_c^2}$ (Collett, 2003), whereas oscillations at the frequency $\Omega$ have almost been suppressed. Only for the case when the frequency $\Omega = 14\pi$ is a multiple of frequency $\omega = 2\pi$, the oscillations at the frequency $\omega$ disappear, and the output power is modulated at frequency $\Omega$. This is like the experimental data shown in Figure 1.21, where HML is stabilized only when $\omega = 2\pi$ , i.e., when the satellites' frequencies are matching the frequency spacing between harmonics. The HML mechanism looks like the vector mode-locking at the fundamental frequency (Sergeyev et al., 2021). By adjusting the in-cavity polarization controller POC2, we were able to increase the circular birefringence strength that leads to the generation of two satellite lines around the q = 0 harmonic frequency. When the birefringence-based modulation frequency $\omega$ approaches the fundamental frequency, the modulation of the harmonic at the frequency $\omega$ disappears, and $TR_{2m}$ is activation results in modulation of q = 0 cavity mode with the frequency of $TR_{2m}$. The amplitude of the output power (Figure 1.32g) along with the Stokes parameters shown was small, and so SOP was locked (Figure 1.32i).

The trajectories shown in Figure 1.32c, f are different from the experimentally observed (Figure 1.21d). However, the DOP = 62% in Figure 1.21d is indication that the SOP evolves at the time scale faster than the polarimeter resolution of 1 μs and, after averaging over ten roundtrips, can merge to the dot at the Poincare sphere (Kbashi et al., 2019b). By using a low-pass filter (Hanning window with the transmission spectrum $T(f) = \left(1 + \cos\left(\pi f / f_c\right)\right)/2$, $f \le f_c = 1 MHz$) we processed time domain waveforms shown in Figure 1.32c and found that circle was transformed to the dot with DOP = 61.7% that is close to the experimental data. The suppression of the oscillations at the frequency $\omega$ indicates that the linewidth is narrowing due to oscillations only at the frequency $\Omega$ (Figure 1.32h).

We highlight that the analysis of HML based on the excitation of $TR_{2m}$ provides just a qualitative approach to the linewidth suppression. The presence of oscillations of the output power at the frequency of $TR_{2m}$ mode and cancelation of the oscillations at the frequency related to the linear and circular birefringence results in narrowing the RF line and increased SNR of 30 dB as shown in Figure 1.22c.

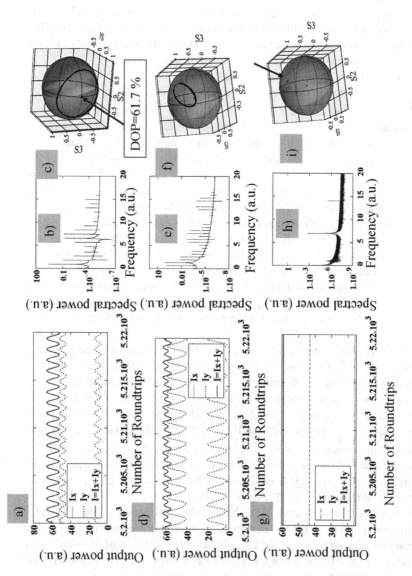

**FIGURE 1.32** Results of the numerical modeling. (a), (d), (g) The output power $I = I_x + I_y$ (solid line) and total power $I = I_x + I_y$ (dots) and total power $I = I_x + I_y$ (solid line); (b), (e), (h) Spectrum of the oscillations; (c), (f), (i) trajectories on the Poincaré sphere. Parameters: time is normalized to the roundtrip time, frequency $\Omega$ – to the fundamental frequency; birefringence strengths $\beta_L, \beta_C$ – to the fiber length; (a)–(i) $\Omega = 7, A_0 = 0.1$; ellipticity of the pump wave $\delta = 0.5$; (a)–(c) $\beta_L = 2\pi / \sqrt{5}$, $\beta_C = 0$; (d)–(f) $\beta_L = 2\pi / \sqrt{5}$, $\beta_C = 2\pi\sqrt{2} / \sqrt{5}$; (g)–(i) $\beta_L = 2\pi / \sqrt{5}$, $\beta_C = 2\pi\sqrt{4} / \sqrt{5}$. The other parameters are found in text (Adapted from Sergeyev et al., 2021).

## 1.16  SELF-PULSING IN FIBER LASERS (THEORY)

Output power self-pulsing (or self-Q-switching, SQS) in fiber lasers at frequencies of 10–100 kHz is a phenomenon resulting in emergence of auto-oscillations without external modulation (Le Boudec et al., 1993; Sanchez et al., 1995; Sanchez and Stephan, 1996; Barmenkov and Kir'yanov, 2004; Kir'yanov et al., 2004; Lee and Agrawal, 2010; Sergeyev et al., 2010; Mallek et al., 2013; Toral-Acosta et al., 2014). To explain variety of self-pulsing operations, the following mechanisms have been suggested: (i) effect of saturable absorber by unpumped section of active fiber (Toral-Acosta et al., 2014) or clustered erbium ions (Le Boudec et al., 1993; Sanchez et al., 1995; Sanchez and Stephan, 1996); (ii) stimulated Brillouin scattering (Mallek et al., 2013), (iii) self-phase modulation (Lee and Agrawal, 2010); (iv) coherence and anti-coherence resonance (CR an ACR) scenario where multimode and polarization instabilities play the role of an external noise source (Sergeyev et al., 2010); (iv) the pump-to-signal intensity noise transfer (PSINT) (Barmenkov and Kir'yanov, 2004); (v) power-dependent thermo-induced lensing in Er-doped fiber (Kir'yanov et al., 2004). The X-ray-absorption fine structure spectroscopy (XAFS) has revealed a short-range coordination order (SRCO) of erbium ions rather than pair clustering (Peters and Houde-Walter, 1998). Though in high concentration LIEKKI TM fibers SRCO is suppressed (Tammela et al., 2003; Sergeyev and Khoptyar, 2007), however, SQS still presents (Sergeyev et al., 2010). It has been also found, that the PSINT can contribute to low-frequency self-pulsing only slightly above the first lasing threshold (Barmenkov and Kir'yanov et al., 2004). Also, coherence and CR and ACR are feasible scenarios but the models of CR and ACR-based self-pulsing have not been developed yet (Sergeyev et al., 2010). Given that the saturable effect of unpumped active fiber, Brillouin scattering, and self-phase modulation have a high threshold, these mechanisms can be feasible only for high power Yb-doped lasers (Lee and Agrawal, 2010; Mallek et al., 2013; Toral-Acosta et al., 2014). Power-dependent thermo-induced lensing can induce low-threshold self-pulsing (Kir'yanov et al., 2004), but can't explain an origin of the experimentally observed complex self-pulsing regimes (Le Boudec et al., 1993; Sanchez et al., 1995).

In this section, we review a new concept of a tunable vector self-pulsing in Er-doped fiber laser developed by (Sergeyev, 2016). The approach is based on Equation (1.43). To reveal self-pulsing without a saturable absorber at frequencies less than the fundamental frequency of mode locking, linear stability along saddle index analysis (Tigan and Opriş, 2008) have been applied to find conditions for emerging complex vector attractors on the Poincare sphere as a function of the laser parameters such as the cavity birefringence, and power and ellipticity of the pump wave. Stability analysis was validated by numerical simulations which demonstrated that double scroll polarization attractor (DSPA) can exist as a result of the polarization symmetry breaking in isotropic cavity without a saturable absorber whereas increased ellipticity of the pump wave and in-cavity birefringence leads to deformation of DSPA to chaotic attractor and further to limit cycle and stable focus. Thus, the obtained theoretical results provides an insight into the experimental data on the complex of self-pulsing regimes including chaos and rogue waves.

To describe the evolution of the laser SOP on the Poincare sphere in terms of the Stokes parameters and the population of the first excited level in $Er^{3+}$ doped active medium we use a simplified form of Equation (1.43):

$$\frac{dS_0}{dt} = \left(\frac{2\alpha_1 f_1}{1+\Delta^2} - 2\alpha_2\right)S_0 + \frac{2\alpha_1 f_2}{1+\Delta^2}S_1 + \frac{2\alpha_1 f_3}{1+\Delta^2}S_2,$$

$$\frac{dS_1}{dt} = \gamma S_2 S_3 + \frac{2\alpha_1 f_2}{1+\Delta^2}S_0 + \left(\frac{2\alpha_1 f_1}{1+\Delta^2} - 2\alpha_2\right)S_1 - \beta_c S_2 - \frac{2\alpha_1 f_3 \Delta}{1+\Delta^2}S_3,$$

$$\frac{dS_2}{dt} = -\gamma S_1 S_3 + \frac{2\alpha_1 f_3}{1+\Delta^2}S_0 + \beta_c S_1 + \left(\frac{2\alpha_1 f_1}{1+\Delta^2} - 2\alpha_2\right)S_2 + \frac{2\alpha_1 f_2}{1+\Delta^2}S_3,$$

$$\frac{dS_3}{dt} = \frac{2\alpha_1 \Delta f_3}{1+\Delta^2}S_1 - \frac{2\alpha_1 f_2}{1+\Delta^2}S_2 + \left(\frac{2\alpha_1 f_2}{1+\Delta^2} - 2\alpha_1\right)S_3,$$

$$\frac{df_1}{dt} = \varepsilon\left[\frac{(\chi_s-1)I_p}{2} - 1 - \left(1 + \frac{I_p\chi_p}{2} + d_1 S_0\right)f_1 - \left(d_1 S_1 + \frac{I_p\chi_p}{2}\frac{(1-\delta^2)}{(1+\delta^2)}\right)f_2 - d_1 S_2 f_3\right],$$

$$\frac{df_2}{dt} = \varepsilon\left[\frac{(1-\delta^2)}{(1+\delta^2)}\frac{I_p(\chi_s-1)}{4} - \left(\frac{I_p\chi_p}{2} + 1 + d_1 S_0\right)f_2 - \left(\frac{(1-\delta^2)}{(1+\delta^2)}\frac{I_p\chi_p}{2} + d_1 S_1\right)\frac{f_1}{2}\right],$$

$$\frac{df_3}{dt} = -\varepsilon\left[\frac{d_1 S_2 f_1}{2} + \left(\frac{I_{p\chi p}}{2} + 1 + d_1 S_0\right)f_3\right].$$

$$(1.45)$$

Next, Equation (1.45) were linearized in the vicinity of the steady state solution ($S_0 \neq 0$, $S_1 = S_2 = 0$, $S_3 = \pm S_0$) and find numerically eigenvalues for the parameters quite close to the experimental ones (Sergeyev et al., 2014): viz. $L = 17$ m, $\alpha_1 = \ln(10)6.4$, $\alpha_4 = \ln(10)0.5$, $\chi = 3/2$, $\Delta = 0.1$, $I_p = 10$, $\gamma = 2 \times 10^{-6}$, $\varepsilon = 10^{-4}$. As a result, we find eigenvalues $\lambda$ and saddle index $\nu$ as follows:

$$\lambda_0 = 0, \quad \lambda_{1,2} = -\gamma_1 \pm i\omega_1, \quad \lambda_{3,4} = -\gamma_2 \pm i\omega_2, \quad \lambda_{5,6} = \rho \pm i\omega_3, \quad \nu = \left|\frac{\gamma_1}{\rho}\right| \quad (1.46)$$

$$(\omega_{1,2,3} \neq 0, \rho, \gamma_1, \gamma_2 > 0, \gamma_1 > \gamma_2).$$

In view of eigenvalues for steady states ($S_0 \neq 0$, $S_1 = S_2 = 0$, $S_3 = S_0$) and ($S_0 \neq 0$, $S_1 = S_2 = 0$, $S_3 = -S_0$) are equal, conditions of the chaos existence in the neighborhood of the heteroclinic orbit can be written in the form of generalized Shilnikov theorem as follows: $\gamma^2 > 0$, $\rho^2 > 0$ and $\nu < 1$ (Tigan and Opriş, 2008).

The saddle index $\nu$ as a function of the anisotropy of the pump $(1-\delta^2)/(1+\delta^2)$ and birefringence strength $\beta$ is shown in Figure 1.33. The internal area bounded by the surface shown in Figure 1.33 and the surface $\nu = 1$ defines the area of chaotic oscillations (Tigan and Opriş, 2008).

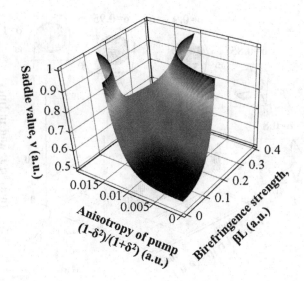

**FIGURE 1.33** Saddle index ($\nu = |\text{Re}\{\lambda_1\}/\text{Re}\{\lambda_5\}|<1$) as a function of the pump power $I_p$, birefringence strength $\beta$ and the anisotropy of pump $(1-\delta^2)/(1+\delta^2)$ in the vicinity of the steady state solution ($S_0 \neq 0$, $S_1 = S_2 = 0$, $S_3 = \pm S_0$). Parameters: $L = 17m$, $\alpha_1 = \ln(10)6.4$, $\alpha_2 = \ln(10)0.5$, $\chi = 3/2$, $\Delta = 0.1$, $\gamma = 2 \times 10^{-6}$, $\varepsilon = 10^{-4}$. The threshold pump power for cw operation $I_{p,th} = 5.123$, the boundaries of self-pulsing for $\beta = 0$, $\delta = 1$: $I_{pmin} = 5.134$, $I_{pmax} = 140$ (Adapted from Sergeyev, 2016).

As follows from Figure 1.33 chaotic oscillations exist for the wide range of the pump power and very narrow range of ellipticity, and the cavity birefringence strength. Tunability of the dynamics with tuning the ellipticity of the pump wave is shown in Figures 1.34 and 1.35. As follows from Figure 1.34, double scroll spiral attractor can exist without a saturable absorber (Figure 1.34, $\delta = 1$) and takes the forms of to the chaotic attractor, limit cycle and stable focus with increased ellipticity of the pump wave. However, with increased the birefringence strength, spiral attractor changes shape to the chaotic attractor and the limit cycle (Figure 1.35). As follows from Figure 1.36 (filled squares), increased birefringence results in modulation of the Stokes parameters with the frequency $(2\beta-\alpha_1 f_2 \Delta/(1+\Delta^2))$ (Sergeyev et al., 2017).

As follows from Figure 1.36, spiral attractor is transforming into the limit cycle for $\beta L > 0.017$ and $\delta = 1$. However, for $\delta = 0.9$ chaotic oscillations of the output power is emerging again as a result of re-activation of oscillations close to the steady state of $S_0 \neq 0$, $S_1 = S_2 = 0$, $S_3 = \pm S_0$ (filled triangles in Figure 1.36) and the presence of birefringence-driven oscillations around steady state $S_0 \neq 0$ $S_1 = \pm S_0$, $S_2 = S_3 = 0$ (filled squares in Figure 1.36). The complexity of demonstrated self-pulsing regimes can be interpreted based on the coupled oscillators theory (Aronson et al., 1990; Pikovsky et al., 2002; Arenas et al., 2008; Thévenin et al., 2011) where the phase difference $\Delta\varphi$ for two orthogonal linearly polarized modes plays the role of relative

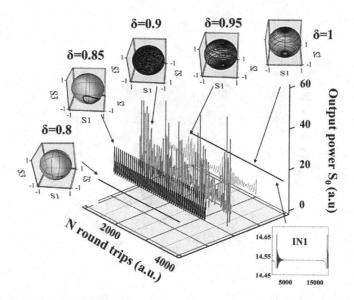

**FIGURE 1.34** The calculated from Equation (1.45) dynamic waveforms and corresponding trajectories on the Poincare sphere as a function of ellipticity of the pump wave $\delta$ ($I_p = 20$, $\beta = 0$). Inset (IN1) shows auto-oscillations of the output power that corresponds to the double scroll polarization attractor. Parameters: $L = 17$ m, $\alpha_1 = \ln(10)6.4$, $\alpha_2 = \ln(10)0.5$, $\chi = 3/2$, $\Delta = 0.1$, $\gamma = 2 \times 10^{-6}$, $\varepsilon = 10^{-4}$ (Reprinted with permission from Sergeyev, 2016).

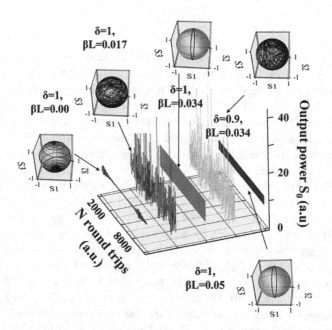

**FIGURE 1.35** The calculated from Equation (1.45) dynamic waveforms and corresponding trajectories on the Poincare sphere as a function of the birefringence strength $\beta$ ($I_p = 20$). Parameters: $L = 17$ m, $\alpha_1 = \ln(10)6.4$, $\alpha_2 = \ln(10)0.5$, $\chi = 3/2$, $\Delta = 0.1$, $\gamma = 2 \times 10^{-6}$, $\varepsilon = 10^{-4}$ (Reprinted with permission from Sergeyev, 2016).

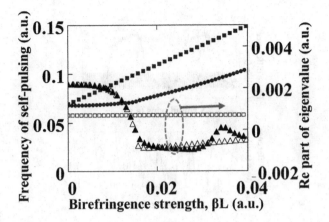

**FIGURE 1.36** Frequency of self-pulsing (in red) and real part of the corresponding eigenvalue (in blue) in the vicinity of steady solution $S_0 \neq 0$, $S_1 = S_2 = 0$, $S_3 = \pm S_0$ (filled circles and empty and filled triangles) and $S_0 \neq 0$ $S_1 = \pm S_0$, $S_2 = S_3 = 0$ (filled squares and empty squares). Parameters: $I_p = 20$, $\delta = 1$ (squares and empty triangles), $\delta = 0.9$ (filled triangles). Real part for eigenvalue in the vicinity of steady state $S_0 \neq 0$ $S_1 = \pm S_0$, $S_2 = S_3 = 0$ (empty squares) is multiplied on $10^{-3}$. All frequencies are normalized to the fundamental frequency which is inversely proportional to the round-trip time. (Reprinted with permission from Sergeyev, 2016.)

phase. The following equation for the phase difference $\Delta\varphi$ can be derived by using Equation (1.45):

$$\frac{d\Delta\varphi}{dt} = \Delta\Omega + K_{NL}\sin^2(\Delta\varphi) + K_s \sin(\Delta\varphi) + K_{as}\cos(\Delta\varphi),$$

$$\Delta\Omega = \frac{2\alpha_1 f_2 \Delta}{1+\Delta^2} - 2\beta, K_{NL} = \frac{\gamma L I_{ss}}{6}(I_y - I_x), K_s = -\frac{2(I_y + I_x)}{\sqrt{I_x I_y}}\frac{\alpha_1 f_3}{(1+\Delta^2)}, \quad (1.47)$$

$$K_{as} = \frac{2(I_y - I_x)}{\sqrt{I_x I_y}}\frac{\alpha_1 f_3 \Delta}{(1+\Delta^2)}.$$

Equation (1.47) is the further generalization of Adler equation (Aronson et al., 1990; Pikovsky et al., 2002; Arenas et al., 2008; Thévenin et al., 2011) based on accounting for asymmetry in the coupling of polarization modes ($K_{as}$ coefficient) and polarization modes' coupling based on the Kerr effects ($K_{NL}$). In addition to this, $I_x$ and $I_y$ are also time-dependent variables.

As follows from the form of Equations (1.47 and 1.45), detuning $\Delta\Omega$ and coupling coefficients $K_s$, $K_{as}$, $K_{NL}$ depend on the laser parameters and so mapping of different synchronization regimes (phase locking, phase entrainment with periodically oscillating phase and phase drifting with chaotically oscillating phase) can be done only numerically by using Equation (1.45) (S. Sergeyev, 2016).

For example, the absence of self-pulsing for high pump power can be explained by the increased coupling strength between polarization modes to allow phase

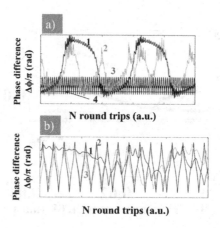

**FIGURE 1.37**   a, b The phase difference evolution. (a) $\beta = 0$ and $\delta = 0.95$ (curve 1), (2) $\delta = 0.9$, (3) $\delta = 0.85$, and (4) $\delta = 0.8$; (b), (1) $\delta = 1$, $\beta = 0.017$; (2) $\delta = 1$, $\beta = 0.034$; (3) $\delta = 0.9$, $\beta = 0.034$. Parameters: $I_p = 20$, $L = 17$m, $\alpha_1 = \ln(10)6.4$, $\alpha_4 = \ln(10)0.5$, $\chi = 3/2$, $\Delta = 0.1$, $\gamma = 2 \times 10^{-6}$, $\varepsilon = 10^{-4}$ (Reprinted with permission from Sergeyev, 2016.)

locking and so achieve a steady-state solution. The results of the phase difference evolution as a function of ellipticity of the pump wave and the birefringence strength are shown in Figure 1.37, b. For $\beta = 0$, it is possible to conclude that detuning is growing faster as compared to the coupling and so phase dynamics changes from the phase entrainment to the phase drift (curves 1 and 2 in Figure 1.37) and chaotic self-pulsing while tuning ellipticity from $\delta = 0.95$ to $\delta = 0.9$. Further decreasing the ellipticity results in the faster growth of coupling that leads to the phase entrainment (curve 3 in Figure 1.37) and periodic power oscillations with the increased frequency and, finally, in phase locked solution (curve 4 in Figure 1.37) and steady-state regime for $\delta = 0.8$. For circularly polarized pump ($\delta = 1$) and increasing birefringence strength, phase difference dynamics is changing from the chaotic drift ($\beta = 0.017$, curve 1 in Figure 1.37b) to the phase entrainment ($\beta = 0.034$, curve 2 in Figure 1.37b) with the fast periodic oscillations of the phase difference and output power correspondingly (Figure 1.34). Aforementioned resonance activation of the chaotic oscillations (Figure 1.34 and 1.35) for ($\delta = 0.9$ and $\beta = 0.034$) results in the chaotic phase drift (curve 3 in Figure 1.37b).

In conclusion, we demonstrate theoretically different vector self-pulsing scenarios in erbium-doped fiber laser. Unlike previous self-pulsing models modulation (Le Boudec et al., 1993; Sanchez et al., 1995; Sanchez and Stephan, 1996; Barmenkov and Kir'yanov, 2004; Kir'yanov et al., 2004; Lee and Agarwal, 2010; Sergeyev et al., 2010; Mallek et al., 2013; Toral-Acosta et al., 2014), we reveal the emergence of the complex vector self-pulsing regimes in terms of the theory of synchronization of coupled oscillators as a transition from the phase locking to new types of phase entrainment and intermittent phase drift regimes as a function of coupling strength and detuning between oscillators' frequencies equation (Aronson et al., 1990; Pikovsky et al., 2002; Arenas et al., 2008; Thévenin et al., 2011).

## ACKNOWLEDGEMENT

The authors would also like to acknowledge support from UK EPSRC project EP/W002868/1, European Union's grants Horizon 2020 ETN MEFISTA (861152) and EID MOCCA (814147).

## REFERENCES

Agrawal, G.P. 2013. *Nonlinear Fiber Optics*. San Diego: Academic Press.

Akhmediev, N., Kibler, B., Baronio, F., Belić, M., Zhong, W.P., Zhang, Y., Chang, W., Soto-Crespo, J.M., Vouzas, P., Grelu, P. and Lecaplain, C. 2016. "Roadmap on optical rogue waves and extreme events." *Journal of Optics* 18(6): 063001.

Arenas, A., Díaz-Guilera, A., Kurths, J., Moreno, Y. and Zhou, C. 2008. "Synchronization in complex networks." *Physics Reports* 469(3): 93–153.

Aronson, D.G., Ermentrout, G.B., and Kopell, N. 1990. "Amplitude response of coupled oscillators." *Physica D* 1990(41): 403–449.

Barmenkov, Y.O. and Kir'yanov, A.V. 2004. "Pump noise as the source of self-modulation and self-pulsing in Erbium fiber laser." *Optics Express*, 12(14): 3171–3177.

Baronio, F., Degasperis, A., Conforti, M. and Wabnitz, S. 2012. "Solutions of the vector nonlinear Schrödinger equations: Evidence for deterministic rogue waves." *Physical Review Letters* 109(4): 044102.

Benjamin, T.B. and Feir, J.E. 1967. "The disintegration of wave trains on deep water Part 1. Theory." *Journal of Fluid Mechanics* 27(3): 417–430.

Boscolo, S., Sergeyev, S.V., Mou, C., Tsatourian, V., Turitsyn, S., Finot, C., Mikhailov, V., Rabin, B. and Westbrook, P.S. 2014. "Nonlinear pulse shaping and polarization dynamics in mode-locked fiber lasers." *International Journal of Modern Physics B*. 28(12): 1442011.

Byrne, J.A., Gabitov, I.R. and Kovačič, G. 2003. "Polarization switching of light interacting with a degenerate two-level optical medium." *Physica D* 2003: 69–92.

Chen, S., Grelu, P. and Soto-Crespo, J.M. 2014. "Dark-and bright-rogue-wave solutions for media with long-wave–short-wave resonance." *Physical Review E* 89(1): 011201.

Chouli, S., and Grelu, P. 2010. "Soliton rains in a fiber laser: An experimental study." *Physical Review A* 81: 063829.

Collett, E. 2003. *Polarized Light in Fiber Optics*. Lincroft, NJ: The polaWave Group.

Cundiff, S.T., Collings, B.C., Akhmediev, N.N., Soto-Crespo, J.M., Bergman, K. and Knox, W.H. 1999. "Observation of polarization-locked vector solitons in an optical fiber." *Physical Review Letters* 82: 3988–3991.

Cundiff, S.T. and Ye, J. 2003. "Colloquium: Femtosecond optical frequency combs." *Reviews of Modern Physics* 75(1): 325.

Desurvire, E. 1994. *Erbium-Doped Fiber Amplifiers, Principles and Applications*. New York: John Wiley & Sons.

Dudley, J.M., Dias, F., Erkintalo, M. and Genty, G. 2014. "Instabilities, breathers and rogue waves in optics." *Nature Photonics* 8(10): 755–764.

Faraday, M. 1831. "On a peculiar class of acoustical figures; and on certain forms assumed by groups of particles upon vibrating elastic surfaces." *Philosophical Transactions of the Royal Society of London* 121: 299–340.

Fontana, F., Begotti, M., Pessina, E.M. and Lugiato, L.A. 1995. "Maxwell-Bloch modelocking instabilities in erbium-doped fiber lasers." *Optics Communications* 114(1–2): 89–94.

Fu, H., and Haken, H. 1987. "Semiclassical dye-laser equations and the unidirectional single-frequency operation." *Physical Review A* 36: 4802–4816.

Geng, Y., Zhou, H., Han, X., Cui, W., Zhang, Q., Liu, B., Deng, G., Zhou, Q. and Qiu, K. 2022. "Coherent optical communications using coherence-cloned Kerr soliton microcombs." *Nature Communications* 13(1): 1–8.

Graham, R. and Haken, H. 1968. "Quantum theory of light propagation in a fluctuating laser-active medium." *Zeitschrift für Physik A Hadrons and Nuclei* 213(5): 420–450.

Grelu, P. and Akhmediev, N. 2012. "Dissipative solitons for mode-locked lasers." *Nature Photonics* 6: 84–92.

Grudinin, A.B. and Gray, S. 1997. "Passive harmonic mode locking in soliton fiber lasers." *JOSA B* 14(1): 144–154.

Gui, L.L., Xiao, X.S., and Yang, C.X. 2013. "Observation of various bound solitons in a carbon-nanotube-based erbium fiber laser." *JOSA B* 30: 158–164.

Haus, J.W., Shaulov, G., Kuzin, E.A. and Sanchez-Mondragon, J. 1999. "Vector soliton fiber lasers." *Optics Letters* 24: 376–378.

Heismann, F. 1994. "Analysis of a reset-free polarization controller for fast automatic polarization stabilization in fiber-optic transmission systems." *Journal of Lightwave Technology* 12(4): 690–699.

Hillerkuss, D., Schmogrow, R., Schellinger, T., Jordan, M., Winter, M., Huber, G.T. et al. 2011. "26 Tbit s-1 21 line-rate super-channel transmission utilizing all-optical fast Fourier transform processing." *Nature Photonics* 5: 364–371.

Jiang, Y., Narushima, T. and Okamoto, H. 2010. "Nonlinear optical effects in trapping nanoparticles with femtosecond pulses." *Nature Physics* 6: 1005–1009.

Kanda, N., Higuchi, T., Shimizu, H., Konishi, K., Yoshioka, K. and Kuwata-Gonokami, M. 2011. "The vectorial control of magnetization by light." *Nature Comunications* 2: 1–5.

Kbashi, H., Sergeyev, S.V., Mou, C., Garcia, A.M., Araimi, M.A., Rozhin, A., Kolpakov, S. and Kalashnikov, V. 2018. "Bright-dark rogue waves." *Annalen der Physik* 530(5): 1700362.

Kbashi, H.J., Sergeyev, S.V., Al Araimi, M., Tarasov, N. and Rozhin, A. 2019a. "Vector soliton rain." *Laser Physics Letters* 16(3): 035103.

Kbashi, H.J., Sergeyev, S.V., Al-Araimi, M., Rozhin, A., Korobko, D. and Fotiadi, A. 2019b. "High-frequency vector harmonic mode locking driven by acoustic resonances." *Optics Letters* 44(21): 5112–5115.

Kbashi, H.J., Zajnulina, M., Martinez, A.G. and Sergeyev, S.V. 2020. "Mulitiscale spatiotemporal structures in mode-locked fiber lasers." *Laser Physics Letters* 035103.

Khanin, Y.A. 2005. *Fundamentals of Laser Dynamics*. Cambridge: Cambridge International Science Publishing.

Kharif, C., Pelinovsky, E. and Slunyaev, A. 2009. *Rogue Waves in the Ocean*. Berlin, Heidelberg: Springer.

Kim, I. and Lee, K.J. 2015. "Axial strain dependence of torsional acousto-optic gratings induced in a form-birefringence optical fiber." *Journal of the Korean Physical Society* 67(3): 465–471.

Kimel, A.V. and Li, M. 2019. "Writing magnetic memory with ultrashort light pulses." *Nature Reviews Materials* 4(3): 189–200.

Kir'yanov, A.V., Il'Ichev, N.N. and Barmenkov, Y.O. 2004. "Excited-state absorption as a source of nonlinear thermo-induced lensing and self-Q-switching in an all-fiber Erbium laser." *Laser Physics Letters* 1(4): 194.

Le Boudec, P., Jaouen, C., Francois, P.L., Bayon, J.F., Sanchez, F., Besnard, P. and Stephan, G. 1993. "Antiphase dynamics and chaos in self-pulsing erbium-doped fiber lasers." *Optics Letters* 18(22): 1890–1892.

Lecaplain, C., Grelu, P. and Wabnitz, S. 2014. "LDynamics of the transition from polarization disorder to antiphase polarization domains in vector fiber lasers." *Physical Review A* 89(6): 063812.

Lee, H. and Agrawal, G.P. 2010. "Impact of self-phase modulation on instabilities in fiber lasers." *IEEE Journal of Quantum Electronics*, 46(12): 1732–1738.

Leners, R., and Stéphan, G. 1995. "Rate equation analysis of a multimode bipolarization Nd3+ doped fibre laser." *Quantum and Semiclassical Optics: Journal of the European Optical Society Part B* 7(5): 757.

Li, F., Wai, P.K.A. and Kutz, J.N. 2010. "Geometrical description of the onset of multi-pulsing in mode-locked laser cavities." *JOSA B* 27(10): 2068–2077.

Li, X.L., Zhang, S.M., Meng, Y.C., Hao, Y.P., Li, H.F. et al. 2012. "Observation of soliton bound states in a graphene mode locked erbium-doped fiber laser." *Laser Physics* 22: 774–777.

Liu, X. and Pang, M. 2019. "Revealing the buildup dynamics of harmonic mode-locking states in ultrafast lasers." *Laser & Photonics Reviews* 13(9): 1800333.

Lugiato, L., Prati, F. and Brambilla, M. 2015. *Nonlinear Optical Systems*. Cambridge, UK. Cambridge University Press.

MacPhail-Bartley, I., Wasserman, W.W., Milner, A.A. and Milner, V. 2020. "Laser control of molecular rotation: Expanding the utility of an optical centrifuge." *Review of Scientific Instruments* 91(4): 045122.

Mallek, D., Kellou, A., Leblond, H. and Sanchez, F. 2013. "Instabilities in high power fiber lasers induced by stimulated Brillouin scattering." *Optics Communications* 308: 130–135.

Mandon, J., Guelachvili, G. and Picqué, N. 2009. "Fourier transform spectroscopy with a laser frequency comb." *Nature Photonics* 25: 99–102.

Misawa, K. 2016. "Applications of polarization-shaped femtosecond laser pulses." *Advances in Physics: X* 1(4): 544–569.

Mou, Ch., Sergeyev, S.V., Rozhin, A. and Turitsyn, S.K. 2011. "All-fiber polarization locked vector soliton laser using carbon nanotubes." *Optics Letters* 36: 3831–3833.

Mou, Ch., Sergeyev, S.V., Rozhin, A.G. and Turitsyn, S.K. 2013. "Bound state vector solitons with locked and precessing states of polarization." *Optics Express* 21(22): 26868–26875.

Niang, A., Amrani, F., Salhi, M., Grelu, P. and Sanchez, F. 2014. "Rains of solitons in a figure-of-eight passively mode-locked fiber laser." *Applied Physics B* 116(3): 771–775.

Onorato, M., Waseda, T., Toffoli, A., Cavaleri, L., Gramstad, O., Janssen, P.A.E.M., Kinoshita, T., Monbaliu, J., Mori, N., Osborne, A.R. and Serio, M. 2009. "Statistical properties of directional ocean waves: The role of the modulational instability in the formation of extreme events." *Physical Review Letters* 102(11): 114502.

Onorato, M., Residori, S., Bortolozzo, U., Montina, A. and Arecchi, F.T. 2013. "Rogue waves and their generating mechanisms in different physical contexts." *Physics Reports* 528(2): 47–89.

Ortac, B., Zavyalov, A., Nielsen, C.K., Egorov, O., Iliew, R. et al. 2010. "Observation of soliton molecules with independently evolving phase in a mode-locked fiber laser." *Optics Letters* 35: 1578–1580.

Ovsyannikov, I.M. and Shilnikov, L.P. 1987. "On systems with a saddle-focus homoclinic curve." *Mathematics of the USSR-Sbornik* 58: 557–574.

Pang, M., Jiang, X., He, W., Wong, G.K.L., Onishchukov, G., Joly, N.Y., Ahmed, G., Menyuk, C.R. and Russell, P. St J. 2015. "Stable subpicosecond soliton fiber laser passively mode-locked by gigahertz acoustic resonance in photonic crystal fiber core." *Optica* 2: 339–342.

Peccianti, M., Pasquazi, A., Park, Y., Little, B.E., Chu, S.T., Moss, D.J. and Morandotti, R. 2012. "Demonstration of a stable ultrafast laser based on a nonlinear microcavity." *Nature Communications* 3(1): 1–6.

Perego, A.M., Tarasov, N., Churkin, D.V., Turitsyn, S.K. and Staliunas, K. 2016. "Pattern generation by dissipative parametric instability." *Physical Review Letters* 116(2): 028701.

Pessina, E.M., Bonfrate, G., Fontana, F. and Lugiato, L.A. 1997. "Experimental observation of the Risken-Nummedal-Graham-Haken multimode laser instability." *Physical Review A* 56(5): 4086.

Pessina, E.M., Prati, F., Redondo, J., Roldán, E. and De Valcarcel, G.J. 1999. "Multimode instability in ring fiber lasers." *Physical Review A* 60(3): 2517.

Peters, P.M. and Houde-Walter, S.N. 1998. "Local structure of Er3+ in multicomponent glasses." *Journal of Non-Crystalline Solids* 239(1–3): 162–169.

Picqué, N. and Hänsch, T.W. 2019. "Frequency comb spectroscopy." *Nature Photonics* 13(3): 146–157.

Pikovsky, A., Rosenblum, M. and Kurths, J. 2002. *Synchronization: A Universal Concept in Nonlinear Science.* Cambridge: Cambridge University Press.

Pilipetskii, A.N., Luchnikov, A.V. and Prokhorov, A.M. 1993. "Soliton pulse long-range interaction in optical fibres: The role of light polarization and fibre geometry." *Soviet Lightwave Communications* 3(1): 29–39.

Pupeza, I., Zhang, C., Högner, M. and Ye, J. 2021. "Extreme-ultraviolet frequency combs for precision metrology and attosecond science." *Nature Photonics* 15(3): 175–186.

Quiroga-Teixeiro, M., Clausen, C.B., Sørensen, M.P., Christiansen, P.L. and Andrekson, P.A. 1998. "Passive mode locking by dissipative four-wave mixing." *JOSA B* 15(4): 1315–1321.

Risken, H. and Nummedal, K. 1968. "Instability of off resonance modes in lasers." *Physics Letters A* 26(7): 275–276.

Rohrmann, P., Hause, A. and Mitschke, F. 2012. "Solitons beyond binary: Possibility of fibre-optic transmission of two bits per clock period." *Scientific Reports* 2: 866.

Sanchez, F., LeFlohic, M., Stephan, G.M., LeBoudec, P. and Francois, P.L. 1995. "Quasi-periodic route to chaos in erbium-doped fiber laser." *IEEE Journal of Quantum Electronics* 31(3): 481–488.

Sanchez, F. and Stephan, G. 1996. "General analysis of instabilities in erbium-doped fiber lasers." *Physical Review E* 53(3): 2110.

Seong, N.H. and Kim, D.Y. 2002. "Experimental observation of stable bound solitons in a figure-eight fiber laser." *Optics Letters* 27: 1321–1323.

Sergeyev, S.V. 1996. "Orientational-relaxation dependent bichromatic operations of a ring cavity dye laser with polarized pumping." *Optics Communications* 131(4–6): 399–407.

Sergeyev, S. 1999. "Spontaneous light polarization symmetry breaking for an anisotropic ring cavity dye laser." *Physical Review A* 59: 3909–3917.

Sergeyev, S. 2003. "Model of high-concentration erbium-doped fibre amplifier: Effects of migration and upconversion processes." *Electronics Letters* 39(6): 511–512.

Sergeyev, S., Popov, S. and Friberg, A.T. 2005. "Influence of the short-range coordination order of erbium ions on excitation migration and upconversion in multicomponent glasses." *Optics Letters* 30(11): 1258–1260.

Sergeyev, S., Popov, S., Khoptyar, D., Friberg, A.T. and Flavin, D. 2006. "Statistical model of migration-assisted upconversion in a high-concentration erbium-doped fiber amplifier." *JOSA B* 23(8): 1540–1543.

Sergeyev, S. and Khoptyar, D. 2007. "Theoretical and experimental study of migration-assisted upconversion in high-concentration erbium doped silica fibers." *Laser Optics: Solid State Lasers and Nonlinear Frequency Conversion.* 151–158.

Sergeyev, S.V., O'Mahoney, K., Popov, S. and Friberg, A. 2010. "Coherence and anticoherence resonance in high-concentration erbium-doped fiber laser." *Optics Letters* 35(22): 3736–3738.

Sergeyev, S.V., Mou, C., Rozhin, A. and Turitsyn, S.K. 2012. "Vector solitons with locked and precessing states of polarization." *Optics Express* 20(24): 27434–27440.

Sergeyev, S.V. 2014. "Fast and slowly evolving vector solitons in mode-locked fibre lasers." *Philosophical Transactions of the Royal Society A: Mathematical, Physical and Engineering Sciences* 372(2027): 20140006.

Sergeyev, S.V., Mou, Ch., Turitsyna, E.G., Rozhin, A., Turitsyn, S.K. and Blow, K. 2014. "Spiral attractor created by vector solitons." *Light: Science & Applications* 3(1): e131–e131.

Sergeyev, S.V. 2016. "Vector self-pulsing in erbium-doped fiber lasers." *Optics Letters* 41(20): 4700–4703.

Sergeyev, S.V., Kbashi, H., Tarasov, N., Loiko, Y. and Kolpakov, S.A. 2017. "Vector-resonance-multimode instability." *Physical Review Letters* 118(3): 033904.

Sergeyev, S., Kbashi, H., Mou, C., Martínez, A., Kolpakov, S. and Kalashnikov, V. 2018. "Vector rogue waves driven by polarisation instabilities." In *Nonlinear Guided Wave Optics: A Testbed for Extreme Waves*, S. Wabnitz, pp. 9.1–9.24. IOP Publishing Ltd.

Sergeyev, S., Kolpakov, S. and Loika, Y. 2021. "Vector harmonic mode-locking by acoustic resonance." *Photonics Research* 9(8): 1432–1438.

Sergeyev, S.V., Eliwa, M. and Kbashi, H. 2022. "Polarization attractors driven by vector soliton rain." *Optics Express* 30(20): 35663–35670.

Shelby, R.M., Levenson, M.D. and Bayer, P.W. 1985. "Guided acoustic-wave Brillouin scattering." *Physical Review B* 31(8): 5244.

Solli, D.R., Ropers, C., Koonath, P. and Jalali, B. 2007. "Optical rogue waves." *Nature* 450(7172): 1054–1057.

Soto-Crespo, J.M., Grelu, Ph., Akhmediev, N., and Devine, N. 2007. "Soliton complexes in dissipative systems: Vibrating, shaking, and mixed soliton pairs." *Physical Review E* 75: 016613.

Spanner, M., Davitt, K.M., and Ivanova, M. Yu. 2001. "Stability of angular confinement and rotational acceleration of a diatomic molecule in an optical centrifuge." *Journal of Chemical Physics* 115: 8403–8410.

Sulimany, K., Lib, O., Masri, G., Klein, A., Fridman, M., Grelu, P., Gat, O. and Steinberg, H. 2018. "Bidirectional soliton rain dynamics induced by casimir-like interactions in a graphene mode-locked fiber laser." *Physical Review Letters* 121(13): 13390.

Sylvestre, T., Coen, S., Emplit, P. and Haelterman, M. 2002. "Self-induced modulational instability laser revisited: Normal dispersion and dark-pulse train generation." *Optics Letters* 27(7): 482–484.

Szwaj, C., Bielawski, S., Derozier, D. and Erneux, T. 1998. "Faraday instability in a multimode laser." *Physical Review Letters* 80: 3968.

Tammela, S., Hotoleanu, M., Kiiveri, P., Valkonen, H., Sarkilahti, S. and Janka, K. 2003. "Very short Er-doped silica glass fiber for L-band amplifiers." In *Optical Fiber Communication Conference*. Optical Society of America, WK3.

Tang, D.Y., Zhang, H., Zhao, L.M. and Wu, X. 2008. "Observation of high-order polarization-locked vector solitons in a fiber laser." *Physical Review Letters* 101: 153904.

Thévenin, J., Romanelli, M., Vallet, M., Brunel, M. and Erneux, T. 2011. "Resonance assisted synchronization of coupled oscillators: Frequency locking without phase locking." *Physical Review Letters*, 107(10): 104101.

Tigan, G. and Opriş, D. 2008. "Analysis of a 3D chaotic system." *Chaos, Solitons & Fractals* 36(5): 1315–1319.

Tlidi, M., Staliunas, K., Panajotov, K., Vladimirov, A.G. and Clerc, M.G. 2014. "Localized structures in dissipative media: From optics to plant ecology." *Philosophical Transactions of the Royal Society A: Mathematical, Physical and Engineering Sciences* 372: 20140101.

Tong, L., Miljković, V.D. and Käll, M. 2010. "Alignment, rotation, and spinning of single plasmonic nanoparticles and nanowires using polarization dependent optical forces." *Nano Letters* 10: 268–273.

Toral-Acosta, D., Martinez-Rios, A., Selvas-Aguilar, R., Kir'yanov, A.V., Anzueto-Sanchez, G. and Duran-Ramirez, V.M. 2014. "Self-pulsing in a large mode area, end-pumped, double-clad ytterbium-doped fiber laser." *Laser Physics* 24(10): 105107.

Tsatourian, V., Sergeyev, S.V., Mou, C., Rozhin, A., Mikhailov, V., Rabin, B., Westbrook, P.S. and Turitsyn, S.K. 2013a. "Polarisation dynamics of vector soliton molecules in mode locked fibre laser." *Scientific Reports* 3(1): 1–8.

Tsatourian, V., Sergeyev, S.V., Mou, C., Rozhin, A., Mikhailov, V., Rabin, B., Westbrook, P.S. and Turitsyn, S.K. 2013b. "Fast polarimetry of multipulse vector soliton operation." In *39th European Conference and Exhibition on Optical Communication (ECOC 2013)*. pp. 1–3.

Turitsyna, E.G., Smirnov, S.V., Sugavanam, S., Tarasov, N., Shu, X., Babin, S.A., Podivilov, E.V., Churkin, D.V., Falkovich, G. and Turitsyn, S.K. 2013. "The laminar–turbulent transition in a fibre laser." *Nature Photonics* 7: 783–786.

Udem, Th., Holzwarth, R. and Hänsch, T.W. 2002. "Optical frequency metrology." *Nature* 416: 233–237.

Varshalovich, D.A., Moskalev, A.N. and Khersonskii, V.K. 1988. *Quantum Theory of Angular Momentum*. Singapore: World Scientific.

Voigt, T., Lenz, M.O., Mitschke, F., Roldán, E. and De Valcarcel, G.J. 2004. "Experimental investigation of Risken–Nummedal–Graham–Haken laser instability in fiber ring lasers." *Applied Physics B* 79(2): 175–183.

Wang, X., Sun, M., Liang, Q., Yang, S., Li, S. and Ning, Q. 2020. "Observation of diverse structural bound-state patterns in a passively mode-locked fiber laser." *Applied Physics Express* 13(2): 022009.

Weill, R., Bekker, A., Smulakovsky, V., Fischer, B. and Gat, O. 2016. "Noise-mediated Casimir-like pulse interaction mechanism in lasers." *Optica* 3(2): 189–192.

Willemsen, M.B., van Exter, M.P. and Woerdman, J.P. 2001. "Polarization loxodrome of a vertical-cavity semiconductor laser." *Optics Communications* 199: 167–173.

Williams, Q.L. and Roy, R. 1996. "Fast polarization dynamics of an erbium-doped fiber ring laser." *Optics Letters* 21: 1478–1480.

Wu, X., Tang, D.Y., Luan, X.N. and Zhang, Q. 2011. "Bound states of solitons in a fiber laser mode locked with carbon nanotube saturable absorber." *Optics Communications* 284: 3615–3618.

Yan, Z.Y. 2010. "Financial rogue waves." *Communications in Theoretical Physics* 54(5): 947.

Zakharov, V.E. and Ostrovsky, L.A. 2009. "Modulation instability: The beginning." *Physica D: Nonlinear Phenomena* 238(5): 540–548.

Zavyalov, A., Iliew, R., Egorov, O., and Lederer, F. 2009. "Dissipative soliton molecules with independently evolving or flipping phases in mode-locked fiber lasers." *Physical Review A* 80: 043829.

Zeghlache, H. and Boulnois, A. 1995. "Polarization instability in lasers. I. Model and steady states of neodymium-doped fiber lasers." *Physical Review A* 52: 4229s–4242.

Zhang, H., Tang, D.Y., Zhao, L.M. and Wu, X. 2009. "Observation of polarization domain wall solitons in weakly birefringent cavity fiber lasers." *Physical Review E* 80: 052302.

Zhao, L.M., Tang, D.Y., Zhang, H. and Wu, X. 2008. "Polarization rotation locking of vector solitons in a fiber ring laser." *Optics Express* 16: 10053–10058.

Zhao, X., Li, T., Liu, Y., Li, Q. and Zheng, Z. 2018. "Polarization-multiplexed, dual-comb all-fiber mode-locked laser." *Photonics Research*, 6(9): 853–857.

# 2 Recent Development of Polarizing Fiber Grating Based Mode-Locked Fiber Laser

*Zinan Huang, Yuze Dai, Qianqian Huang,
Zhikun Xing, Lilong Dai, Weixi Li, Zhijun Yan and
Chengbo Mou*

## CONTENTS

2.1 Introduction .................................................................................................69
2.2 Principle, Fabrication and Characterization of the 45° TFG .........................72
   2.2.1 Principle of the 45° TFG ....................................................................72
   2.2.2 Fabrication and Characterization of the 45° TFG ...............................73
      2.2.2.1 UV Inscription .....................................................................73
      2.2.2.2 Femtosecond Laser Inscription ...........................................74
2.3 Mode-Locked Fiber Lasers Based on 45° TFG .............................................75
   2.3.1 Influence of 45° TFG Performance on Mode-Locked
      Fiber Laser ..........................................................................................76
   2.3.2 Stretched-Pulse Mode-Locked Fiber Laser .........................................77
   2.3.3 Wavelength Tunable/Switchable and Multi-Wavelength
      Mode-Locked Fiber Laser ...................................................................79
   2.3.4 Pulse State Switchable Mode-Locked Fiber Laser ..............................85
   2.3.5 GHz Harmonic Mode-Locked Fiber Laser .........................................88
   2.3.6 Mode-Locked Laser Using a 45° TFG as an In-Fiber
      Polarization Beam Splitter ..................................................................89
   2.3.7 Mode-Locked Fiber Laser Based on Femtosecond Laser
      Inscribed 45° TFG ..............................................................................93
2.4 Conclusions and Perspectives ......................................................................95
Acknowledgement ................................................................................................96
References .............................................................................................................96

## 2.1 INTRODUCTION

Polarization is one of the most important properties of light, and understanding and manipulating light polarization is critical in various optical applications, such as optical sensing (Enokihara et al., 1987), optical fiber communications (Han and Li,

DOI: 10.1201/9781003206767-2

2005), radar (Villarini and Krajewski, 2010), laser (Zhao et al., 2008) and biological microscopy (Brasselet, 2011). A polarizer is the key optical element in controlling the polarization, which transmits a desired polarization state while reflecting, absorbing or deviating the rest. A wide variety of popular polarizer designs are available to date, including glass/plastic linear polarizers (Pei et al., 2015), thin film polarizers (Dobrowolski and Waldorf, 1981), wire grid polarizers (Yamada et al., 2009) and polarizing plate/cube beam splitters (Lopez and Craighead, 1998). However, most of them are based on conventional free-space components, which limits further miniaturization and integration to the modern optical fiber systems while introducing additional coupling loss and maintenance costs. Researchers and engineers have begun to explore and develop the in-fiber polarization dependent device to enable all-fiber systems with low loss. great compatibility, high reliability and low cost.

One important application of in-fiber polarizing devices is mode-locked fiber lasers. Mode-locked fiber lasers have emerged as powerful platforms for fundamental science as well as industrial and medical applications with unique advantages, such as compact structure, excellent beam quality, high efficiency, low maintenance and low costs (Kim and Song, 2016; Peng et al., 2018). A simple and reliable passive mode-locking technique for the implementation of an ultrafast fiber laser with desirable performances remains a subject of intense interest to researchers in the field. As a key device for the achievement of passive mode-locking, the invention of a saturable absorber (SA) based on semiconductor nonlinear optical absorption has led to the development of novel nanomaterial nonlinear optics in the last few decades (Gomes et al., 2004; Lecourt et al., 2012; Gao et al., 2021). However, the material type SA still has some problems, including a complicated preparation process, low damage threshold and degradation with long-term use. Another ideal way to generate ultrashort pulses is to utilize techniques based on the intrinsic Kerr nonlinearity of optical fibers including nonlinear polarization evolution (NPE) (Tamura, Haus, and Ippen, 1992), a nonlinear amplifying loop mirror (NALM) (Fermann et al., 1990) and a nonlinear optical loop mirror (NOLM) (Doran and Wood, 1988). Of these, the NPE method with intra-cavity polarizing elements has been extensively used due to the benefits of ultrashort response time, high power tolerance and large modulation depth (Li et al., 2015). It is well known that the performance of the polarizing element, which plays the key role of NPE mode-locking, is closely related to the quality of output ultrashort pulse.

As early as 1980, Eickhoff presented an in-line fiber-optic polarizer through partially grinding off the cladding on one side of a single-mode fiber (SMF) and replacing it with metal thin film (Eickhoff, 1980). Subsequently, the technique of polishing the fiber laterally and coating a birefringent crystal or a metal film technique has been widely used as a simple and effective fiber-processing method for the production of in-fiber polarizers (Wang et al., 2020). In 2011, graphene was first utilized as a replacement of conventional metal layer on top of a D-shaped fiber (Bao et al., 2011). Such types of polarizers also feature a broadband response. A single-polarization fiber was reported in 1983, using a combination of high stress-induced birefringence with a depressed or W-type cladding structure that can be used as a fiber polarizer (Simpson et al., 1983). While maintaining its perfectly all-fiber format, such polarizer fibers would normally require a considerable length to achieve the desired polarization extinction ratio. In 1990, Peng et al. demonstrated a fiber-polarization beam splitting scheme using twin-elliptic core fibers (Peng, Tjugiarto, and Chu, 1990). Later, fiber polarizers that take advantage of the unique structure of photonic crystal fiber (PCF) were presented (Li and Xiao, 2019). And in recent years, microfibers with strong evanescent field are also utilized to develop the polarizer or the polarization beam splitter (PBS) to control the state of the polarization (Zhang et al., 2015a; Zhou et al., 2021).

In addition, another influential in-fiber polarization-dependent device is implemented based on a fiber-grating structure, in a process which is known as polarizing fiber grating. In contrast to the polarization-dependent elements based on side-polished fibers, polarizing fibers and microfibers, the fabrication of polarizing fiber grating does not require delicate and sophisticated manipulation, the performance is more reliable and the structure is more robust. Compared with the PCF-based polarizers, polarizing fiber gratings can be adapted to most types of fiber, resulting in lower costs and avoiding the large splice loss of PCF. Consequently, the polarizing fiber grating is an ideal type of in-fiber polarizer. To date, polarizing fiber gratings can be categorized into long period grating (LPG) (Kurkov et al., 1997; Ortega et al., 1997; Bachim and Gaylord, 2003), chiral fiber grating (Du, Shu, and Zuowei, 2016; Xue et al., 2022) and 45° tilted fiber gratings (45° TFGs) (Zhou et al., 2005; Yan et al., 2011; Posner et al., 2019). In 1997, Kurkov et al. first demonstrated a long period grating (LPG) written on a polarization-maintaining (PM) fiber that exhibits polarization splitting based on polarization mode dispersion (Kurkov et al., 1997). Immediately afterward, a high-performance in-fiber polarizer based on LPG (Ortega et al., 1997) was proved by Ortega et al. A helical LPG inscribed in PM fiber by $CO_2$ laser was presented in 2019 to show excellent polarization-dependent characteristic with its polarization-dependent loss (PDL) exceeding 30 dB (Jiang et al., 2019). Chiral fiber gratings with double helix symmetry were introduced in 2004 (Kopp et al., 2004) with polarization-selective transmission and thus can be used as polarizers. Last but not least, driven by continuous improvements in grating inscription technology and demands for widespread applications, the fiber Bragg grating with a tilt angle of 45° was developed; this achieved a polarization extinction ratio (PDL) higher than 33 dB over a 100 nm operation range (Zhou et al., 2005). To date, these polarizing fiber gratings have been extensively studied and applied in optical fiber sensors, optical wireless communications and fiber lasers.

The typical schematic of NPE mode-locked fiber laser using a polarizing fiber grating is shown in Figure 2.1. Although mode-locked fiber lasers based on LPGs (Huang et al., 2019a) and chiral fiber gratings (Du, Shu, and Zuowei, 2016) have been proven, the narrow bandwidth of LPGs and the fabrication complexity of chiral fiber gratings limit their extensive applications in mode-locked fiber lasers. The 45° TFG with excellent polarization dependent characteristics, broadband response, high compatibility and the simple preparation method, has been utilized to implement diversified mode-locked fiber lasers with various operation regimes and wavebands (Yan et al., 2016).

In this chapter, we summarize the recent development of mode-locked fiber lasers based on polarizing fiber gratings, and in particular 45° TFGs. The first section provides a background introduction on polarizing fiber gratings and highlights their application in mode-locked fiber lasers. The second section briefly reviews the

---

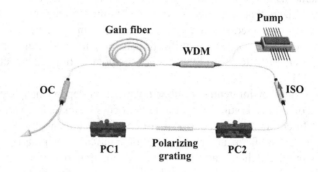

**FIGURE 2.1**  The typical schematic diagram of mode-locked fiber laser based on a polarizing fiber grating.

principle, two major fabrication techniques and the characterization of the 45° TFG. The third section reviews in detail the latest developments in various mode-locked fiber lasers based on 45° TFG. Finally, the chapter contains a summary and an assessment of the future prospects for polarizing fiber grating-based mode-locked fiber lasers.

## 2.2   PRINCIPLE, FABRICATION AND CHARACTERIZATION OF THE 45° TFG

### 2.2.1   PRINCIPLE OF THE 45° TFG

A 45° TFG utilizes Brewster's law for the transmission of specific linear polarization state (Yan et al., 2011). Brewster's angle claims that the angle of incidence at which the reflected beam is only $s$-polarized and the transmitted beam becomes partially $p$-polarized. According to Snell's law, the magnitude of Brewster's angle is the inverse tangent of the ratio of the refractive indices of two media. Since the refractive index change caused by fiber-grating inscription is typically of the order of $10^{-4}$, Brewster's angle of the fiber grating is approximately 45°. That is to say, the 45° TFG is actually an equivalent set of "pile-of-plates" arrangement at Brewster's angle, as schematically shown in Figure 2.2. When unpolarized light is transmitted through 45° TFG, the $s$-polarization light is coupled from the forward-propagating core mode into radiation mode and tapped out from the side of grating, while the residual $p$-polarization light propagates along the fiber. After the refractive index modulation structures, a degree of polarization $p$-polarization light is finally obtained at the other end of the 45° TFG. One needs to notice that from the transmission perspective, the 45° TFG is a bidirectional polarizing device.

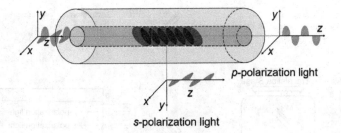

**FIGURE 2.2**   Principle of 45° TFG as in-fiber polarizer.

## 2.2.2   FABRICATION AND CHARACTERIZATION OF THE 45° TFG

### 2.2.2.1   UV Inscription

The development of the phase mask scanning method has greatly improved the fiber-grating inscription process. Hence the fabrication of 45° TFG using such a method has been well established. A typical schematic of the 45° TFG inscription using phase mask scanning technique with UV irradiation is displayed in Figure 2.3a (Yan et al., 2016). In distinction to the standard fiber grating inscription, the mask-scanned UV laser beam is de-focused by a cylindrical lens onto the fiber core. Before the inscription, the standard SMF is required to be hydrogen-loaded to enhance the photosensitivity. Surely, commercially available or home-made intrinsic photosensitive fibers are also preferred for the inscription of 45° TFG. It should be noted that using a custom phase mask with a 33.7° tilted pattern or rotating a normal phase mask by 33.7° externally with respect to the fiber axis ensures that the grating pattern has a 45° tilt angle.

A typical microscopic image of the UV-inscribed 45° TFG with the phase mask scanning mask technique is shown in Figure 2.3b. Moreover, the key parameter of the 45° TFG as a polarization-dependent device is its PDL or polarization extinction ratio (PER), which is defined as the maximum (peak-to-peak) difference in transmission as the input polarization varies over all its states. The PDL of 45° TFG is linearly proportional to the grating length and the square of the refractive index modulation. Figure 2.3c shows the polarization characteristics of a typical UV-inscribed 45° TFG evaluated by a commercial fiber vector analyzer (LUNA, OVA5000). It can be seen that the PDL of UV-inscribed 45° TFG can reach 50 dB nowadays. The appearance of small ripples in the spectrum may be well explained by the back-and-forth reflection of the radiation modes back at the cladding/air boundary, thereby forming the cladding mode oscillations. According to previous studies, this oscillation has no observable effect on the laser mode-locking performance and can be eliminated by immersing the grating in a medium with refractive index similar to the fiber cladding (Mou et al., 2009). It can be well understood from which the visibility of such oscillation is normally quite low. Surely, by properly designing the extrinsic environmental condition through specific coating or refractive index matching, the visibility of the oscillation can be very much enhanced. Despite the measured spectral range being limited by the light source, the PDL consistently maintains high value in the available range, indicating excellent broadband performance.

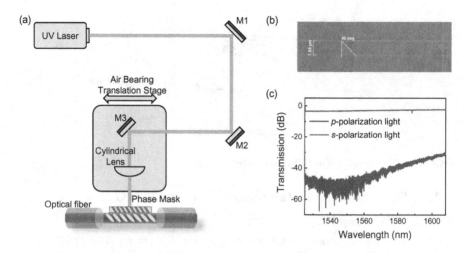

**FIGURE 2.3** (a) Typical 45° TFG inscription system using the phase-mask scanning technique with UV irradiation. (b) The microscope image of the grating structure of a typical UV inscribed 45°-TFG in SMF-28 fiber. (c) Transmission of a typical UV inscribed 45° TFG measured at two orthogonal polarization states.

### 2.2.2.2   Femtosecond Laser Inscription

Compared with the traditional grating inscription technique with UV irradiation, the fabrication technique of grating that uses femtosecond laser has received considerable attention in the last two decades owing to its unique advantages, including fiber materials' photosensitivity independency, and the simplicity of the process (Bernier et al., 2007; Bernier et al., 2012; He et al., 2021). In 2017, Ioannou et al. proposed a plane-by-plane direct-writing method by a femtosecond laser (Ioannou et al., 2017), which brought new vitality to the development of 45° TFG. As shown in Figure 2.4a, the fiber samples are mounted on a controllable air-bearing translation stage without removing the fiber coating. The femtosecond laser beam is focused inside the fiber by a long working distance objective. As can clearly be seen, the inscription process no longer requires the phase mask and photosensitive enhancement, while maintaining the strength and physical integrity of the optical fiber. In addition, this plane-by-plane femtosecond laser inscription method allows flexible control of the grating period and spatial dimensions (depth, width, angle and length of grating) and avoids the difficulty of aligning the fiber core when using the point-by-point and line-by-line techniques (Theodosiou et al., 2016). The low production complexity and short cycle time are undoubtedly superior on the way to practical use.

The typical microscope image and polarization characteristics of the femtosecond laser-inscribed 45° TFG with plane-by-plane direct writing technique are shown in Figures 2.4b and c. It has to be acknowledged that the large refractive index change through femtosecond laser inscription increases the insertion loss while ensuring sufficient PDL. One likely explanation is that the localization and inhomogeneous of femtosecond laser-induced grating structure and the tilted grating plane lead to strong scattering loss (Williams et al., 2013). In order to guarantee its polarization-selection function, the insertion loss and PDL are compromised by optimizing the pulse energy

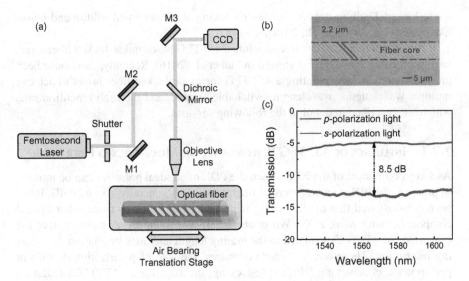

**FIGURE 2.4** (a) Typical 45° TFG inscription system using plane-by-plane direct write method with femtosecond laser. (b) The microscope image of the grating structure of a typical femtosecond laser inscribed 45° TFG in SMF-28 fiber. (c) Transmission of a typical femtosecond laser inscribed 45° TFG measured at two orthogonal polarization states.

of femtosecond laser and grating length. Also, it is noticed that compared with the UV-inscribed 45° TFG, the transmission spectrum of *s*-polarization light is more of a plateau, and there is no ripple caused by the refractive index mismatch owing to the absence of fiber coating. Efforts are underway to optimize the femtosecond laser inscription process to ensure the high PDL value while minimizing the insertion loss.

## 2.3 MODE-LOCKED FIBER LASERS BASED ON 45° TFG

Since the first demonstration of a passively mode-locked Er-doped fiber laser using a 45° TFG as an in-line polarization element in 2010 (Mou et al., 2010), various types of 45° TFG-based fiber lasers with different pulse-shaping mechanisms at different wavebands have been extensively studied. In the pursuit of narrow pulse duration, sub-100 fs dissipative stretch-pulse delivered from a mode-locked Er-doped fiber laser with a 45° TFG were demonstrated (Zhang et al., 2013). And in the pursuit of a high repetition rate, a 251.3 MHz fundamental repetition rate Er-doped fiber ring laser operating in dissipative soliton regime using a 45° TFG was achieved (Zhang et al., 2015b). In addition to Er-doped fiber laser operating in the C-band, the 45° TFG also has a wealth of developments in ultrafast laser applications in the wavelength ranges of 1 μm–2 μm. Liu et al. first reported the use of 45° TFG for an all-normal dispersion Yb-doped fiber laser to achieve mode-locking at 1 μm region (Liu et al., 2012). Immediately afterwards, the bound state pulse evolution (Liu et al., 2013a) and dual-wavelength operation (Liu et al., 2015) were reported successively in 45° TFG based on Yb-doped fiber laser. Meanwhile, Li et al. experimentally demonstrated the first inscription of 45° TFG at 2 μm waveband and employed it into a mode-locked Thulium-doped fiber laser working at conventional soliton and noise-like pulses regimes (Li et al., 2014).

It is worth noting that the aforementioned 45° TFG based mode-locked fiber lasers have been comprehensively reviewed in Yan et al. (2016). Recently, there have been more advances in incorporating a 45° TFG for mode-locked fiber lasers to achieve multiple wavelengths, wavelength switchable/tunable, and the high repetition rate, which are described in detail in the following sessions.

### 2.3.1 INFLUENCE OF 45° TFG PERFORMANCE ON MODE-LOCKED FIBER LASER

As a key component of the NPE effect, the PDL of an ideal polarizer can be infinite. In practice, the PDL of a commercial fiber polarizer is generally above 20 dB. It has been demonstrated that mode-locking is possible in fiber lasers using other optical components with weak PDL (Wu et al., 2010). We found experimentally that the PDL value of 45° TFG is related to the grating length and laser irradiation dose during inscription. Therefore, to better understand the role of polarization-dependent performance in achieving NPE mode-locking, investigation of 45° TFGs PDLs on the performance of mode-locked conventional soliton fiber lasers has been carried out (Wang et al., 2018b).

As shown in Figure 2.5a, four UV-inscribed 45° TFGs with PDL of 5.4 dB, 9 dB, 14 dB and 24 dB at ~1,555 nm were prepared at the same UV irradiation intensity and scan length with different scanning speeds. Their insertion losses were controlled to be ~4 dB, including the loss of both s- and p-light, as shown in Figure 2.5b. Then a typical all-anomalous-dispersion Er-doped fiber laser based on the 45° TFG was built. By replacing the 45° TFGs with different PDLs while keeping the cavity length and pump power almost constant, four mode-locked output states were obtained through adjusting the polarization controllers (PCs). Figure 2.5c~f shows the recorded optical spectra and corresponding autocorrelation (AC) traces of the mode-locked fiber laser based on these four 45° TFGs with different PDLs under the pump power of 160 mW. The spectral bandwidths of the typical soliton spectra are 2.5 nm, 2.8 nm, 4.1 nm and 4.4 nm for PDLs of 5.4 dB, 9 dB, 14 dB and 24 dB, respectively. The slight difference in central wavelength may be caused by the variation of the intra-cavity nonlinearity due to the PC states. The AC traces fit well with a secant hyperbolic profile, then the pulse durations are 1.5 ps, 1.3 ps, 860 fs and 830 fs, respectively. Besides, the mode-locking threshold gradually decreases from 310 mW to 220 mW as the PDL improves from 5.4 dB to 24 dB, while the conversion efficiency increases from 3.9% to 8.1%. It also experimentally demonstrated that the variation of PDL value has no significant effect on the time-bandwidth product (TBP), signal-to-noise ratio (SNR), calculated time jitter and energy fluctuations of the output pulse.

Overall, the experimental results show that with the increase of PDL of 45° TFG, mode-locked pulses with shorter pulse duration, lower mode-locking threshold and higher conversion efficiency can be generated under the conventional soliton regime. This work clarifies the polarization-dependent properties of a 45° TFG affect the NPE mode-locking performance, which has important reference significance for the design of mode-locked fiber lasers using 45° TFG.

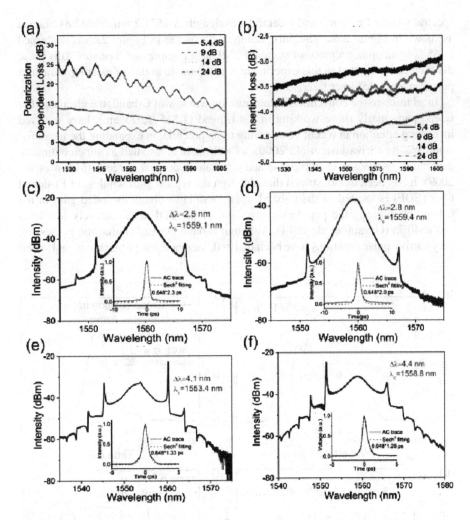

**FIGURE 2.5** (a) The PDL profiles and (b) Insertion loss spectra of four 45° TFG from 1,525 nm to 1,608 nm. Optical spectra and AC traces (inset) of mode-locking operation at the pump power of 160 mW based on four different 45° TFGs with PDL of (c) 5 dB (d) 9 dB (e) 14 dB and (f) 24 dB, respectively. Reprinted with permission from (Wang et al., 2018b) ©2017 IEEE.

## 2.3.2 Stretched-Pulse Mode-Locked Fiber Laser

Although sub-100 fs pulses have been achieved in Er-doped fiber laser using a 45° TFG with a PDL of 30 dB (Zhang et al., 2013), the quest for shorter pulse duration continues building on the work presented in the previous section. Wang et al. employed a 45° TFG with 35 dB PDL to construct a 65 fs mode-locked Er-doped fiber laser with optimized intra-cavity dispersion (Wang et al., 2018a). Mode-locking was easily initialized by adjusting the PCs when the pump power was over 100 mW. When the pump power was further increased to 550 mW, a stretched-pulse with a spectral width of 62.6 nm and a central wavelength at 1571.2 nm could be obtained as shown in Figure 2.6a. The pulse duration illustrated in Figure 2.6b is dechirped to 65 fs by adopting a piece of standard SMF as a compressor. The sub-70 fs pulse duration demonstrated in this experiment is comparable to that of existing ultrashort pulses in Er-doped fiber laser.

In addition to the conventional transmission window of C-band, the ultrafast fiber lasers, particularly those working in the L-band (1,565–1,625 nm), have received increasing attention in recent years owing to its potential in expanding the transmission capacity (Srivastava et al., 2000), as well as in areas such as supercontinuum generation (Kawagoe et al., 2014) and edematous corneal surgery (Morin et al., 2009). It has been demonstrated that although the typical gain window of Er-doped fiber (EDF) is located in the C-band, it is possible to obtain extended gain in the L-band by limiting the population inversion of the EDF to a relatively low level (30%–40%) (Guesmi et al., 2014). Although L-band DS and similariton possessing very narrow pulse durations have been realized, there are few publications on L-band

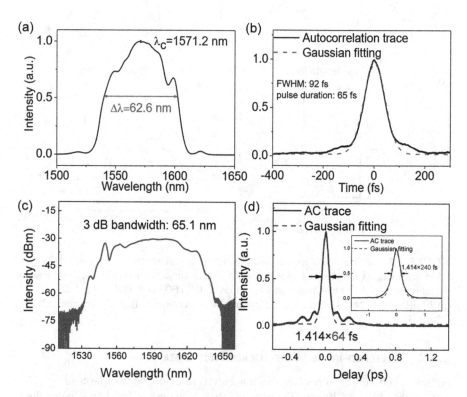

**FIGURE 2.6** (a) Linear optical spectrum and (b) AC trace and its Gaussian fitting of the 65 fs stretched-pulse mode-locked fiber laser based on a 45° TFG. (c) Optical spectrum and (d) AC traces of the compressed and direct output (inset) pulses, and corresponding Gaussian fitting of the L-band 64 fs stretched-pulse mode-locked laser based on a 45° TFG. (Reprinted with permission from (Wang et al. 2018a) ©2018 The Author(s). Reprinted with permission from (Huang et al., 2021a) ©2021 Optical Society of America.)

stretched-pulse all-fiber laser with a pulse duration under 100 fs. Inspired by the results of the above study, a 64 fs L-band stretched-pulse generation from an all-fiber Er-doped laser based on a 45° TFG optimized for the L-band has been presented recently (Huang et al., 2021a). The basic configuration of the laser differs from Wang's work in the choice of a longer piece of highly doped EDF as the gain medium. This provides intra-cavity dispersion compensation near zero while forcing the laser emitting in the L-band, as well as in the use of a 45° TFG in this cavity with a PDL of up to 48 dB at 1,598 nm. As shown in Figure 2.6c, by optimizing the length of the SMF in the cavity, a stretched-pulse operating in the L-band with a 3 dB spectral bandwidth of 65.1 nm was finally achieved. The dechirped pulse with a pulse duration of 64 fs was obtained as depicted in Figure 2.6d. Furthermore, a numerical simulation of the pulse propagation in the cavity has been carried out to confirm the generation of the stretched-pulse in good agreement with the experimental results. In contrast to the previously reported ultrafast fiber laser sources with a pulse duration below 100 fs in the L-band [10, 11], such a laser utilizes only one section of gain fiber, unidirectional pumping and a L-band optimized 45° TFG for which simplifies the cavity design and is more cost-effective in practical applications.

### 2.3.3 Wavelength Tunable/Switchable and Multi-Wavelength Mode-Locked Fiber Laser

For some applications, it is required that the wavelength of a mode-locked laser can be tuned or switched within a certain range. In particular, the availability of a mode-locked fiber laser with a tunable or switchable wavelength is highly relied upon in the fields of broadband multichannel optical communications, fiber sensing, high-capacity optical transmission, wavelength-dependent measurements and optical signal processing techniques (Bewersdorf and Hell, 1998; Vogler et al., 2015; Fu et al., 2018). Generally, the wavelength tuning operation can be performed by varying the effective gain profile. One approach is to control the rare-earth doping concentration of the gain fiber (Wagener et al., 1993), the length of gain fiber (Dong et al., 2005), intra-cavity loss (Zhao et al., 2011) and the pumping scheme (Lin and Lin, 2014), which leads to changes in the population inversion level of the gain fiber and affects the intrinsic gain spectrum. Another more straightforward approach is to insert a tunable spectral filter into the laser cavity, such as mechanical tunable filter, the Mach-Zehnder interferometer (MZI) (Zhang et al., 2011), the Fabry-Perot interferometer (FPI) (Li and Chan, 1999), chirped fiber Bragg grating (He, Liu, and Wang, 2012), long period fiber grating (Wang et al., 2015) and the birefringent comb filter (Luo et al., 2010). Of these, the birefringence comb filter, also known as the Lyot filter, can be formed by the inherent fiber birefringent effect. This not only avoids the large insertion loss of the bulky-type filter, but also has a large tunable range and low cost, which has been widely used in all-fiber lasers.

To obtain the intra-cavity birefringent filtering effect, a pair of polarizing elements and a birefringence medium are usually required. Thanks to the ring configuration of the fiber laser cavity, such a filtering effect can be produced with only one single polarizer in addition to the single-mode fiber applied inside the cavity. Utilizing such a filter, Zou et al. explored the effect of PDL of 45° TFG on the wavelength tuning range in a fiber laser mode-locked by carbon nanotubes (CNTs) (Zou et al., 2018). A piece of CNT-PVA composite film was sandwiched between two FC/PC fiber connectors as the mode-locker. Two 45° TFGs with PDLs of 8 dB (Figure 2.7a) and 19 dB (Figure 2.7c) at 1,550 nm acted as linear polarizers to form the birefringent filter, respectively. Figure 2.7b presents the continuously tuning range of 4.6 nm from 1559.85 nm to 1564.46 nm when using the weak 45° TFG with a PDL of 8 dB. And, as shown in Figure 2.7d, when a strong 45° TFG with a PDL of 19 dB is used, the central wavelength can be easily tuned continuously, from 1553.37 nm to 1568.63 nm, and the tuning range is dramatically increased to 15.26 nm. Benefitting from this study, Lu et al. achieved a wide wavelength tuning range of 36 nm from 1543.49 nm to 1579.08 nm in a CNT-PVA mode-locked fiber laser (Lu et al., 2019). The tunable operation is implemented using a strong 45° TFG with a PDL of 29 dB at 1,550 nm and an 8 cm section of PM fiber to enhance intra-cavity birefringence. Figures 2.7e and f illustrate the PDL response of the 45° TFG and the tunable optical spectra of the Er-doped fiber laser, respectively.

It should be noted that the aforementioned reports are all based on conventional soliton (CS) wavelength tunable mode-locked fiber laser. Nevertheless, CSs generated in the anomalous dispersion cavity are more susceptible to breaking up with the accumulation of excessive nonlinear phase shift, which limits the pulse energy and poses a challenge for practical use. Dissipative solitons (DSs), which are typically produced in a large net-normal-dispersion cavity, can increase their pulse energy by one to two orders of magnitude. The formation of DSs is the result of the combined effect of gain, loss, dispersion, and nonlinearity (Grelu and Akhmediev, 2012), while the spectral amplitude modulation introduced by the spectral filtering effect is also critical (Chong et al., 2006). It is important to point out the aforementioned Lyot filter consisting of a polarizer and a birefringent medium can also be deemed as an effective NPE mode-locked structure, which is very favorable for the direct formation of wavelength tunable DSs. By carefully adjusting the pump power and polarization states in the cavity, it is possible to control comb filtering and NPE effects at the same time. Huang et al. chose a 45° TFG with a PDL of 48.4 dB at 1,593 nm combined with the birefringence of the rest fiber to achieve an NPE mode-locking while creating a comb filtering effect (Huang et al., 2019b). The PDL response of the 45° TFG is exhibited in Figure 2.8a. Figure 2.8b shows the dissipative soliton wavelength tunable operation that was obtained by simply rotating the PCs carefully while keeping the pump power constant. The central wavelength can be shifted from 1,567 nm to 1,606 nm with a range of 39 nm, which is noteworthy as the widest tuning range of DS fiber lasers operating in the communication band.

In addition to the Er-doped fiber laser, with the advantages of high optical-to-optical conversion efficiency, low quantum defect, relative broad gain bandwidth and low thermal effect, the Yb-doped fiber laser operating in the 1 μm region exhibits extensive applications in areas requiring high power output. A pump-controlled wavelength switchable mode-locked Yb-doped fiber laser was developed (Lin et al., 2020), which adopted a similar structure to the aforementioned Er-doped fiber laser. By only raising the pump power, the operating wavelength of DSs can be switched from 1046.51 nm to 1067.9 nm without adjusting the orientation of PCs, as shown in Figure 2.9. The wavelength interval is ~22 nm corresponding to the filtering

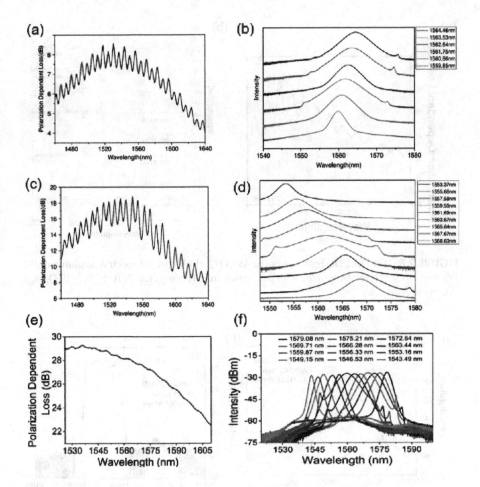

**FIGURE 2.7** The PDL response (a) of the 45° TFG with PDL of 8 dB and the laser spectral tunability (b) based on it. The PDL response (c) of the 45° TFG with PDL of 19 dB and the laser spectral tunability (d) based on it. The PDL response (e) of the 45° TFG with PDL of 29 dB and the laser spectral tunability (f) based on it and 8 cm PM fiber. (Reprinted with permission from (Zou et al., 2018) ©2017 Elsevier B.V.. Reprinted with permission from (Lu et al., 2019) ©2019 IEEE.)

bandwidth of the birefringent comb filter. The wavelength switchable behavior stems from the variation of transmission peak position of the birefringent filter. The pulse peak power increases with an enhancement in pump power. This implies an increase in the nonlinear phase shift, which leads directly to a drift in the transmission peak of the birefringent comb filter and thus to a switch in the laser operating wavelength. Compared to polarization-adjusted wavelength switchable fiber lasers, the pump-adjusted mechanism offers more precision and controllability.

Multi-wavelength mode-locked fiber lasers are important dense wavelength division multiplexing (DWDM) systems because of their excellent spectral selectivity and the ability to provide multiple channels simultaneously (Liu et al., 2013b).

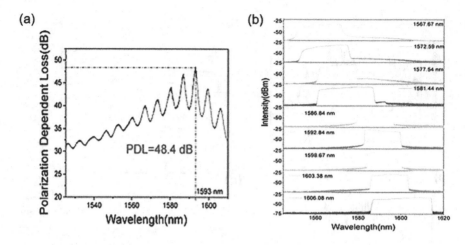

**FIGURE 2.8** (a) The PDL response of the 45° TFG. (b) Measured spectral tunability of the DS mode-locked laser. (Reprinted with permission from (Huang et al., 2019b) © 2019 Optical Society of America.)

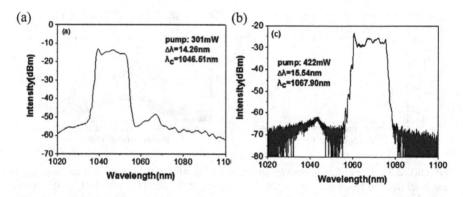

**FIGURE 2.9** (a) Optical spectrum under the pump power of 301 mW and (b) Optical spectrum under the pump power of 422 mW. (Reprinted with permission from (Lin et al., 2020) © 2020 Elsevier GmbH.)

Similar to achieving wavelength tunable/switchable, the key element to generating ultrashort pulses for multi-wavelength excitation also lies in the spectral filtering effect and the saturable absorber. As mentioned above, the 45° TFG combining with the intra-cavity birefringence can serve as both the comb filter and the NPE mode locked, allowing for multi-wavelength mode-locked operation without adding the complexity of the laser cavity. However, since the transmission spectrum of such a birefringent filter is closely related to polarization and the birefringence of SMF cannot be accurately adjusted, the resulting comb filter is highly random in terms of both intensity and wavelength. Undoubtedly, such an all-SMF design limits the number of excitation wavelengths and difficult to precisely control the output wavelength.

To address this issue, a compact all-fiber Lyot filter (AFLF) designed by sandwiching a section of PM fiber with two 45° TFGs inscribed in PM fiber was proposed in 2012 (Yan et al., 2012). Figure 2.10a shows the configuration of just such an AFLF (Li et al., 2019). The structure of the two identical 45° TFGs is inscribed in the PM fiber along the fast or slow axis by UV inscription with the phase-mask scanning technique. A piece of PM fiber is used as a strongly birefringent medium with its fast or slow axis aligned at 45° to that of the two PM 45° TFGs. It should be noted that the fast or slow axis of the two front and rear 45° TFGs are in parallel in the structure. Concretely, when the linear polarized light after the first 45° TFG enters the PM fiber, it is decomposed into two orthogonally polarized beams with equal intensity traveling along the fast and slow axis of the PM fiber, respectively. By virtue of the birefringence of the PM fiber, the two beams experience different phase delay and then interfere with each other when they are combined in the second 45° TFG. That is to say, the relative phase differences are converted into amplitude modulation resulting in comb filtering. The filter bandwidth and free spectral range (FSR) can be easily customized by changing the length of the PM fiber.

Zou et al. experimentally demonstrated a compact multi-wavelength mode-locked Er-doped fiber laser based on such an AFLF, delivering five lasing wavelengths simultaneously (Zou et al., 2020). The AFLF is composed of two 45° TFG with PDL greater than 28 dB from 1,500 nm to 1,600 nm and a 0.8 m section of PM fiber. As shown in Figure 2.10b, the FSR and the bandwidth of the Lyot filter are 7.5 nm and 3.6 nm, respectively. The output spectrum is illustrated in Figure 2.10c with central wavelengths of 1532.78 nm, 1540.64 nm, 1547.76 nm, 1555.41 nm, and 1562.72 nm, respectively. It can be seen the wavelength spacing and bandwidth match well with the transmission spectrum of the filter. Likewise, a switchable multi-wavelength mode-locked Yb-doped fiber laser operating in the 1 μm waveband based on the same filter structure was realized in 2021 (Lin et al. 2021a). The FSR of the filter is 10.2 nm, determined by the 24 cm long PMF. Up to six-wavelength mode-locked pulses can be simultaneously obtained in the experiment as depicted in Figure 2.11a. Moreover, the developed Yb-doped fiber laser also permits the generation of single wavelength switchable DS via finely rotating the PCs. Specifically, Figure 2.11b exhibits the central wavelengths of the five pulses are located at 1035.26 nm, 1044.93 nm, 1055.62 nm, 1066.11 nm and 1076.63 nm, respectively.

It is well known as the formation of CSs is typically accompanied by sharp Kelly sidebands due to the constructive interference between the soliton and the co-propagating dispersive wave (Du et al., 2019). On the one hand, the Kelly sidebands are much wider than the femtosecond pulse in the time domain, resulting in the inter-pulse interactions that reduce the mode-locking stability and increase the bit error rate in communication (Dennis and Duling, 1994). On the other hand, the Kelly sidebands have a higher gain coefficient than the main lobe of the pulse during the amplification, which limits the amplification efficiency of CSs. Therefore, it is necessary to suppress the Kelly sidebands in some applications. In addition to achieving multi-wavelength mode-locked lasers, the aforementioned compact AFLF is an ideal choice to effectively restrain the strong Kelly sidebands of CSs benefitting from the controllable FSR and bandwidth and the large filtering depth (Li et al., 2019). An AFLF with an FSR of 20.8 nm and depth of 9 dB was prepared with two PM 45°

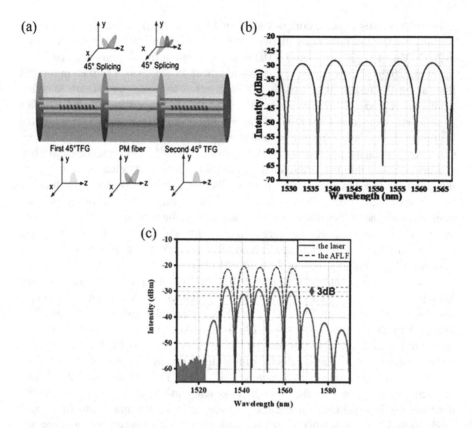

**FIGURE 2.10**  (a) Configuration of an AFLF based on two 45° TFGs and a PM fiber cavity. (b) Measured optical transmission spectrum of the AFLF. (c) The mode-locked optical spectrum (continuous line) under the pump power of 273 mW and the optical spectrum (dotted line) of the AFLF. (Reprinted with permission from (Li et al., 2019) ©2019 Chinese Laser Press, Reprinted with permission from (Zou et al., 2020) ©2020 IEEE.)

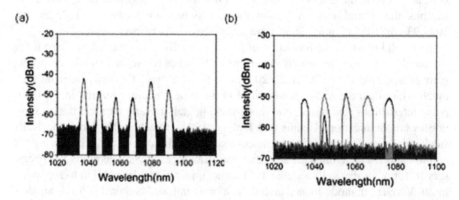

**FIGURE 2.11**  (a) The mode-locked optical spectrum under the pump power of 430 mW. (b) Wavelength switching of single pulse DS. (Reprinted with permission from (Lin et al. 2021a) © 2021 The Author(s).)

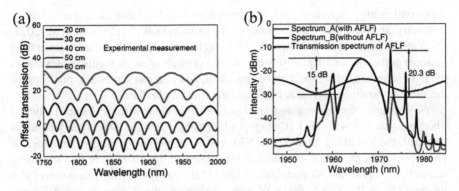

**FIGURE 2.12** (a) Optical transmission spectrum of the AFLF using PMFs with different lengths. (b) The mode-locked optical spectrum with/without the AFLF under the pump power of 1.55 W. (Reprinted with permission from (Li et al., 2019) © 2019 Chinese Laser Press.)

TFG centered at ~1,950 nm and a piece of PM fiber. The filter with optimized FSR was employed in an all-fiber hybrid mode-locked Tm-doped laser to achieve effective Kelly sidebands suppression. Figure 2.12a and b shows the transmission spectra of the Lyot filters with different lengths of PM fiber and the suppression effect of Kelly sidebands, respectively. The effect of different filter transmission spectra on the laser output performance was also explored. A longer PM fiber leads to a smaller filter bandwidth, which introduces additional chirp by narrowing the spectral bandwidth of the mode-locked pulse. On the contrary, a shorter PM fiber results in a larger filter bandwidth, which has a limited effect on the Kelly sidebands suppression. Therefore, the best sidebands suppression can be obtained by designing highest transmission wavelength of the filter coincide with the central wavelength of the generated soliton, and the two adjacent lowest transmission wavelengths fall at the symmetric first-order Kelly sidebands.

### 2.3.4 PULSE STATE SWITCHABLE MODE-LOCKED FIBER LASER

It is well known that the interaction between linear effects and nonlinear effects in fibers, especially dispersion distribution, self-phase modulation, cross-phase modulation and filtering effect, enables various pulse-shaping mechanisms in the passively mode-locked fiber laser. So far, several major pulse-shaping mechanisms, including CS, stretched-pulse, DS have been successively implemented in mode-locked fiber lasers using a 45° TFG. Apart from the single-pulse states, a variety of multipulse states, including bound state, HML and noise-like pulses (NLP), have also been observed. Each pulse state shows different evolution characteristics and output properties. For example, DSs formed under balanced normal dispersion, nonlinear effects, gain and loss are highly tolerant to the accumulation of nonlinear phase shifts. As a consequence, DS mode-locking becomes a preferred choice for average power and pulse energy scaling in fiber lasers. Another important example is the NLPs that can be generated under different dispersion conditions have a broad and smooth optical spectrum and low coherence, and are therefore very promising for supercontinuum generation, optical sensing based on low-coherence spectral inter-ferometry, and optical coherence tomography (Suzuki et al., 2015; Dong et al., 2021; Zhao et al., 2021). The development of a single laser cavity capable of generating multiple pulse states with precisely controllable switching of such pulse states makes the light source more compact and economical, and also can serves as a platform for the study of nonlinear dynamic behaviors in ultrafast fiber lasers.

We first report the implementation of a DS-NLP switchable mode-locked Er-doped fiber laser with a 45° TFG (Cheng et al., 2019). After the DS mode-locking is obtained, the switching from DS to NLP can be achieved by simply changing the pump power. In the follow-up study, it was observed that there were five switching stages between DS and NLP regimes (Cheng et al., 2020) when the pump power was increased from 200 mW to 660 mW without adjusting the PCs, as shown in Figures 13a and b. That is to say, the pulse-state switching effect can be precisely controlled by monotonically varying the pump power. In particular, the entire switching process is repeatable and reversible. According to the previous subsection, we know that the NPE structure can be served not only as a SA for mode-locking, but also as a spectral

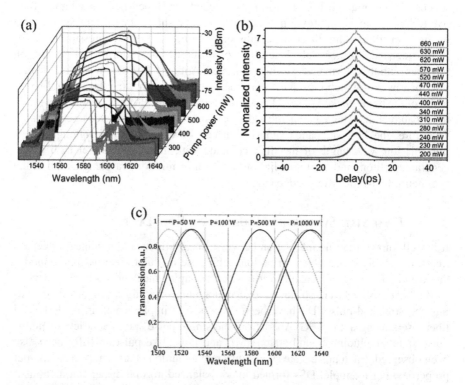

**FIGURE 2.13** (a) Optical spectra and (b) Corresponding AC traces evolution from the DS-NLP multiple switchable fiber laser by increasing pump power under the fixed orientations of PCs. (c) The simulated NPR transmission spectra under different instantaneous power $P = 50$ W, 100 W, 500 W, and 1,000 W. (Reprinted with permission from (Cheng et al., 2020) ©2020 IEEE.)

filter to promote the pulse evolution. Therefore, a simulation of the transmission spectral of the NPE mode-locking mechanism at different pump powers was performed to further investigate the mechanism of such a multi-switching phenomenon of pulse states in the same laser cavity. The typical transmission function of NPE can be described as follows:

$$\left|T_{NPR}\right|^2 = \sin^2(\theta)\sin^2(\psi) + \cos^2(\theta)\cos^2(\psi) + \frac{1}{2}\sin(2\theta)\sin(2\psi)\cos(\Delta_{\varphi L} + \Delta_{\varphi NL}),$$

In this equation, $\theta$ and $\psi$ present the azimuth angles between the polarization directions of 45° TFG and the fast axis of the fiber. $\Delta\varphi_L$ is the linear phase shift, which can be described as $\Delta\varphi_L = 2\pi L \Delta n / \lambda$. And $\Delta\varphi_{NL}$ is the nonlinear phase shift expressed as $\Delta\varphi_{NL} = 2\pi n_2 PL\cos(2\theta)/\lambda A_{eff}$. Here, $\Delta n$, $n_2$, $\lambda$, $L$, $P$, $A_{eff}$ are the fiber birefringence, nonlinear-index coefficient, operating wavelength, cavity length, instantaneous power of the pulse and effective fiber mode field area, respectively. As shown in Figure 2.13c, the polarization states ($\theta$ and $\psi$) were set appropriately so that the effect of nonlinear phase shift on the NPE transfer function could be observed by varying only the pulse instantaneous power. The sinusoidal transmission spectrum implies the existence of two different states of positive and negative feedback for the NPE mechanism. Since NLP is a wave packet containing multiple sub-pulses with high peak power, its nonlinear phase shift is the accumulation of multiple sub-pulses within the wave packet, which is much larger than that of a DS. As a result, a smaller $P$ value $(P = 50\,\text{W})$ is used to represent the NPE transmission profile in the DS regime, while a larger $P$ value $(P = 1000\,\text{W})$ represents the NPE transmission profile in the NLP regime. It can be seen that the increase in pump power leads to a transmittance increase at 1,580 nm (the central wavelength in the experiment), which is in a positive feedback mechanism. When the pump power continues to rise until the peak power clamping effect occurs, the DSs switch to NLP, which has been reported in many previous studies. In the meantime, as the pump power increases, the gain spectrum becomes stronger and wider, thus causing it to blue shift. At this point, the NPE operates in a negative feedback mechanism, which means that the higher the instantaneous power, the lower the NPE transmission. In that case, the NPE transmission spectrum leaps back to its initial state and reaches a new equilibrium. Accordingly, the laser returns to the DS mode-locked state. Eventually, the two switching processes are repeated in the cavity until the pump power increases to the maximum value. They conclude that such a multi-shuttle behavior between two distinct pulse state can be attributed to the interaction of the transmission of NPE mechanism and the gain filtering effect.

On the basis of these findings, a pump-controlled pulse state switchable Yb-doped mode-locked fiber laser based on a AFLF was demonstrated (Lin et al., 2021b). As mentioned, the AFLF incorporated here consists of a pair of PM 45° TFGs with a section of PM fiber at a specific splicing angle. As shown in Figure 2.14, the pulse state of the laser recognizes switching from DS to NLP and then to DS by only continuously increasing the pump power from 177 mW to 691 mW. It is worth mentioning that the DSs after the pump power reaches 323 mW all appear in the bound state due to the enhanced self-phase modulation.

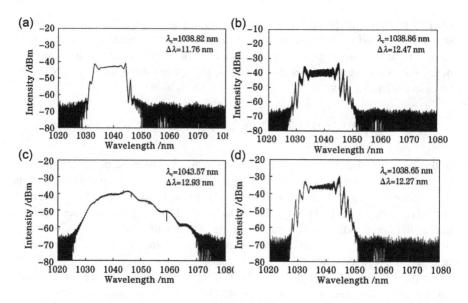

**FIGURE 2.14** Optical spectra of the pulse state switchable mode-locked Yb-doped fiber laser at (a)177 mW. (b) 323 mW. (c) 455 mW. (d) 691 mW. (Reprinted with permission from (Lin et al. 2021b) ©Chinese Laser Press.)

Overall, this kind of pulse state switchable fiber laser precisely controlled by pump power, makes it feasible to develop a compact and genuine multi-functional light source with high controllability. Moreover, the results give a further insight into the pulse-shaping mechanisms and provide an excellent research platform for further exploration of the nonlinear phenomena in fiber lasers.

## 2.3.5 GHz Harmonic Mode-Locked Fiber Laser

Besides pursuing the narrow pulse duration and high output power, the repetition rate, as one of the most important indications in evaluating the performances of the mode-locked fiber laser, has been demanded by various applications, including high-volume optical communication, high-speed optical sampling and optical frequency metrology (Schlager, Hale, and Franzen, 1993; Haus and Wong, 1996). Although an active mode-locked technique makes it simple to extend the repetition rate of the pulses to the order of GHz, the external modulator is bulky and expensive, and its limited response time makes it challenging to obtain femtosecond pulses. In a passively mode-locked fiber laser, high-repetition rate pulses can be achieved by shortening the cavity length (Cheng et al., 2017) or harmonic mode locking (HML) (Wu and Dutta, 2000). Using the former idea, Zhang et al. implemented a 250 MHz fundamental repetition rare mode-locked Er-doped fiber laser based on a 45° TFG (Zhang et al., 2015b). In spite of the in-fiber 45° TFG enabling very integrated and compact laser configuration, reaching GHz-level pulse output is very challenging due to the sophisticated requirements for both gain fiber and nonlinearity for mode-locking in a short cavity.

Alternatively, HML as a distinctive case of multi-pulse operation, where the split pulses owing to peak power limiting and energy quantization effects are arranged at uniform intervals and amplitude. Consequently, HML can be removed from the limitations of cavity length and additional modulators, allowing the pulse repetition rate to grow to multi-GHz at multiples of the fundamental repetition rate. In 2019, Ling et al. demonstrated a GHz L-band femtosecond HML Er-doped fiber laser using an L-band optimized 45° TFG (Ling et al., 2019). Typically, the 45° TFG with PDL greater than 30 dB at the L-band region served as a polarizer in the NPE mode-locking structure, sandwiched between two PCs. A stable single-soliton mode-locking operation was achieved by gradually increasing the pump power to 47 mW, and when the pump power was further increased to 218 mW, HML appeared under appropriate polarization states. Continuing to raise the pump power to the maximum available power of 712 mW without adjusting the PCs anymore, the pulse repetition rate of 4.22 GHz was achieved with the super-mode suppression ratio (SMSR) of 32 dB. The characteristics of output pulses are illustrated in Figure 15a~d, where it can be seen that the pulse quality remains high with a central wavelength of 1581.7 nm and a pulse duration of 810 fs. It should be noted that the continuous wave component located on the optical spectrum has been proved to be an important contributor to stable HML. Moreover, higher harmonic orders can be obtained by finely optimizing the orientations of PCs, they finally extended pulse repetition rate to 7.41 GHz with 20.7 dB SMSR.

In this experiment, there was an interesting and repeatable phenomenon worth noting. It is commonly assumed that the birefringence variation brought by the in-line PC is difficult to accurately predict, so the corresponding variation of harmonic order in HML is also random and irregular. Here, however, it could be observed that the harmonic order changes monotonically from 113th to 39th by unidirectional rotating the orientation of a single PC at a fixed pump power (Ling et al., 2020). As shown in Figure 2.15e, the corresponding repetition rate gradually decreased from 2.21 GHz to 0.77 GHz. In the meantime, the SMSR at each repetition rate was superior to 29 dB. A simulation of the transmission spectra of NPE mechanism under specific conditions was carried out to analyze the observed phenomenon. The NPE transmission function used for the simulation is similar to the one in the previous subsection. The simulation results indicated that the peak of transmission varies linearly with the azimuth angle between the polarization direction of 45° TFG and the fast axis of the fiber under specific conditions, and this azimuth angle represented the rotation of the PC. Therefore, it is speculated that the unidirectional rotating of the PC causes a monotonic variation in the peak of the NPE transmission spectrum, which affects the energy used for HML, thus achieving a monotonic decrease in the harmonic order. Such an intra-cavity birefringence-controlled HML fiber laser with adjustable repetition rate in the GHz range has considerable potential for applications in photonic switching, all-optical sampling, high-speed optical communication, *etc.*

### 2.3.6    Mode-Locked Laser Using a 45° TFG as an In-Fiber Polarization Beam Splitter

As has been shown above, the grating with a tilt angle of 45° can directly and effectively compel *s*-polarized light out of the grating while allowing *p*-polarized light

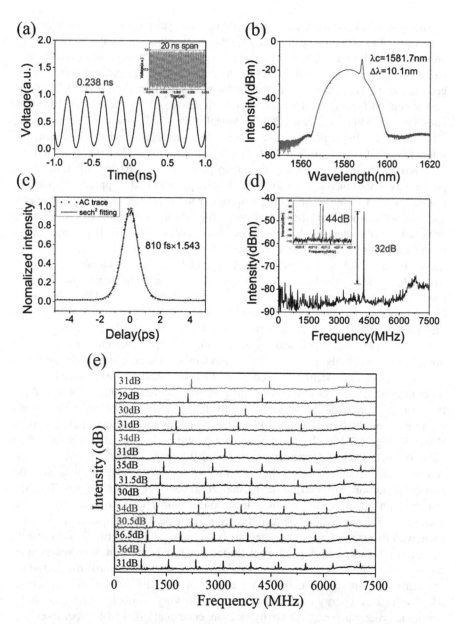

**FIGURE 2.15** (a) Pulse trace. (b) Optical spectrum. (c) AC trace and (d) RF spectra in 7.5 GHz span and inset: RF spectrum of the 4.22 GHz HML pulses. (e) RF spectra with the repetition rate of 2.21, 2.11, 1.87, 1.77, 1.68, 1.58, 1.41, 1.31, 1.26, 1.21, 1.11, 0.95, 0.86, and 0.77 GHz. (Reprinted with permission from (Ling et al., 2019) © The author(s). Reprinted with permission from (Ling et al., 2020) © 2020 Optical Society of America.)

to propagate through the fiber core. Most of the literature on the 45° TFG based mode-locked fiber laser has focused on the utilization of its transmission light, i.e., the *p*-polarized light that remains only in the fiber core after multiple refractions and reflections. However, the s-polarized light which is tapped from the side of the grating as the radiation mode of 45° TFG also has excellent polarization-dependent characteristics. In 2019, Qin et al. systematically analyzed the properties of radiation mode of 45° TFG in theory and practice (Qin et al., 2019). Both simulated and experimental results indicate that the *s*-polarized radiation mode of 45° TFG has an attenuated distribution along the grating axis and a quasi-Gaussian distribution along the fiber radial direction. It is generally understood that the polarization characteristics of the transmission light of the depends on its PDL; the larger the PDL, the closer the transmission light is to fully p-polarized light. In contrast, the polarization distributions of the radiation mode reveals that the radiation light is almost a perfectly linearly polarized light, in which the degree of polarization (DOP) is about 99.886% and is independent of the magnitude of the PDL.

Taking advantage of the unique characteristics of the 45° TFG radiation mode, Xing et al. demonstrated a picosecond mode-locked Er-doped fiber laser using a 45° TFG as an in-fiber polarization beam splitter (Xing et al., 2020). When the incident light to 45° TFG is unpolarized light, that is, considered a rapidly varying random combination of half *p*- and half *s*-polarized light, the intensity of the transmission and radiation light can be expressed by the following equation:

$$I_T = \frac{1}{2}I_{In} + \frac{1}{2}I_{In} \cdot 10^{-\frac{PER}{10}}$$

$$I_R = \frac{1}{2}I_{In} - \frac{1}{2}I_{In} \cdot 10^{-\frac{PER}{10}}$$

where $I_{In}$, $I_T$ and $I_R$ refer to the intensity of incident light, transmission light and radiation light, respectively. Hence, the beam splitting ratio of transmission light and radiation light can be given as:

$$\frac{I_T}{I_R} = \frac{1+10^{-\frac{PER}{10}}}{1-10^{-\frac{PER}{10}}}$$

It can be seen from the equation that by designing the PER of the 45° TFG during the inscription process it is possible to obtain PBS with different beam splitting ratios. Both theoretical calculation and experimental measurement show that with the increase of PER, the intensity of transmission and radiation light gradually tends to be equal. According to the previous subsection, a higher PER is desirable for mode-locking, whereas a larger PER of 45° TFG means a longer grating length, which complicates the coupling due to irregular beam shape resulting from the lateral diffraction. Consequently, a 45° TFG with a grating length of only 5 mm for a PER of 4 dB was prepared as the in-fiber PBS for mode-locking, when almost 70% s-polarized light is radiated out from the side of the grating. The experimental setup of the proposed mode-locked Er-doped fiber laser is illustrated in Figure 2.16a. Unlike the typical 45° TFG based mode-locked fiber laser, the output coupler is no longer required and the 45° TFG forms an artificial SA while also enabling laser output. The radiation light is coupled into the fiber through two cylindrical lenses and

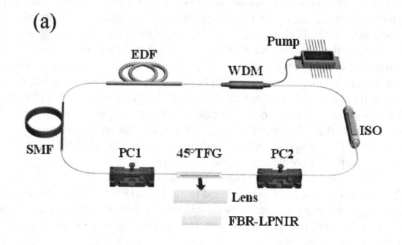

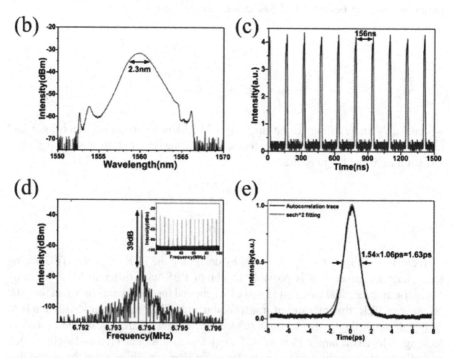

**FIGURE 2.16** (a) Schematic diagram of the mode-locked laser using a 45° TFG based PBS. (b) Optical spectrum. (c) Pulse train. (d) RF spectra with scanning range of 5 kHz and 100 MHz (inset) and (e) AC trace with sech² fitting. (Reprinted with permission from (Xing et al., 2020) © 2020 IEEE.)

a fiber collimator to characterize the output pulse. In practice, the direct use of the radiation light can be achieved through spatial shaping. Figure 2.16b–e presents the measured results of typical CS mode-locked pulses. Additionally, the DOP of the radiation light can be calculated to be 99.8%, which means that the output light consists almost entirely of single linear polarization. This work employs a 45° TFG as a PBS to achieve single polarization output of mode-locked fiber lasers, which lays the groundwork for future development of in-fiber PBS extensions to other wavebands where fiber PBS is expensive or unavailable.

### 2.3.7  Mode-Locked Fiber Laser Based on Femtosecond Laser Inscribed 45° TFG

In the above subsections, the 45° TFGs used in the lasers are all fabricated by a phase-mask scanning technique with UV irradiation. Undoubtedly, the art of UV inscribed 45° TFG is mature enough to easily obtain PDL over 30 dB at specific wavelength. With the processing advantages of femtosecond lasers gradually playing an important role in fiber grating inscription, it has become an irresistible trend to explore advanced inscription techniques of 45° TFGs and applications in mode-locked fiber lasers. The specific approach of plane-by-plane direct-write inscription using a femtosecond laser has been described in detail in the previous section, so only its application to mode-locked fiber lasers is given in the following paragraphs.

As with the UV inscribed 45° TFG, using a femtosecond laser inscribed 45° TFG as an effective in-fiber polarizer, the generation of CSs from an all-anomalous-dispersion mode-locked Er-doped fiber laser was first realized in 2019 (Huang et al. 2019c). The femtosecond laser inscribed 45° TFG with a PDL of 9.5 dB and an insertion loss of 7 dB at 1,560 nm is shown in Figure 2.17a. The two isolators employed in the cavity is intended to reduce the scattering on light transmission due to the femtosecond laser inscription. The performances of CS mode-locking are exhibited in Figure 2.17b and c. The transfer efficiency of the laser was also measured to be 7.72%. Compared with the mode-locked output characteristics reported in (Wang et al., 2018b) using a UV-inscribed 45° TFG with a PDL value of 9 dB, the laser has a comparable mode-locked threshold but a higher conversion efficiency due to the higher output ratio coupler. There is no lower mode-locked threshold because of the higher loss. It needs to be emphasized that such a novel type 45° TFG has more advantages in preparation and application, which improves the integration and reliability of the laser. In addition to CS, the laser also can operate in an NLP regime with proper polarization adjustment.

In quick succession, the application of a femtosecond laser direct-inscribed 45° TFG in a net-normal-dispersion mode-locked Er-doped fiber laser has also been investigated (Huang et al. 2021b). The gain in the cavity is provided by a 2.65 m-long piece of highly-doped Er fiber with a group velocity dispersion of + 61.2 $ps^2$/km, giving a normal net intra-cavity dispersion and allowing for DS generation. In addition, such a long highly-doped gain fiber is chosen to compensate for the high loss caused by the femtosecond laser-inscribed 45° TFG, thus providing sufficient gain and nonlinear accumulation for mode-locking. The single-pulse DS output characteristics are demonstrated in Figure 2.18. The approximately rectangular spectrum

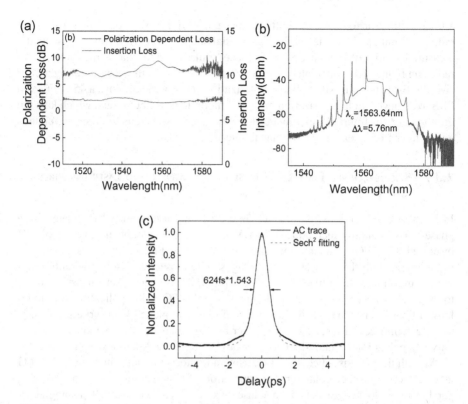

**FIGURE 2.17**   (a) The PDL response and insertion loss of the 45°-TFG. Single soliton pulse characteristics of the mode-locked laser under the pump power of 150 mW. (b) Optical spectrum and (c) AC trace with sech$^2$ fitting. (Reprinted with permission from (Huang et al. 2019) © 2019 Optical Society of America.)

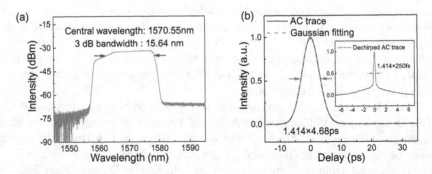

**FIGURE 2.18**   DS pulse characteristics of the mode-locked laser under the pump power of 250 mW: (a) Optical spectrum. (b) AC trace and the corresponding Gaussian fitting. Inset: the dechirped AC trace. (Reprinted with permission from (Huang et al. 2021b) © 2021 Elsevier Ltd.)

profile with steep rising and falling edges indicate the DS mode-locking. Using a section of SMF as an external compressor, the dechirped pulse duration is 250 fs. Moreover, the dual-pulse bound-state and harmonic mode-locking are also observed by increasing the pump power and optimizing the polarization states.

In the meantime, using such a 45° TFG-inscribed via direct-write plane-by-plane femtosecond laser method, an all-fiber mode-locked Thulium-doped fiber laser was also proved in 2019 (Kanagaraj et al., 2019). The PDL of a femtosecond laser-inscribed 45° TFG at 1,870 nm is about 3.5 dB, but in the range beyond 200 nm it is more than 3 dB, which is weak but sufficient to achieve mode-locking. A stable soliton centered at 1,870 nm with a pulse duration of 862 fs was generated at a fundamental repetition rate of 22.34 MHz.

The above results indicate the feasibility and versatility of the femtosecond laser direct-inscribed 45° TFG as a polarization selection device to achieve a NPE mode-locked all-fiber laser. Certainly, the high insertion loss limits the practical laser applications. However, we believe that with the optimization of the inscription process, this combination of flexibility and robustness of a femtosecond laser-inscribed 45° TFG paves a new way for the development of a compact, high-reliability and high-energy all-fiber ultrafast laser.

## 2.4   CONCLUSIONS AND PERSPECTIVES

In conclusion, 45° TFG has proved to be an effective in-fiber polarizing element for NPE mode-locked fiber lasers. It has been successfully implemented in fiber lasers working at various wavelength regions under a number of operation regimes. Surely, for lasers at other wavelength regions which has not been experimentally demonstrated, the grating devices would be feasible. In particular, inscribing such fiber-grating structures at short or mid-infrared wavelength range can be challenging considering the availability of grating fabrication facilities for these ranges. Although TFG inscription in ZBLAN fiber has been reported, the polarization-dependent property is not good enough to demonstrate mode-locking so far., mode-locking have not been demonstrated by far.

Nevertheless, the advantages of such polarizing grating devices in mode-locked fiber lasers have not been fully revealed. For example, such grating structure could be formed in different types of specialty optical fibers such as multimode fiber, high nonlinear fiber, tapered fiber, and active fiber, *etc*. This would contribute to a more advanced mode-locked fiber laser with a highly compact or robust configuration. In addition, the 45° TFG-based mode-locked fiber lasers are now mainly focused on seed lasers only. With proper amplifier design, one could achieve high-power ultrafast fiber laser. Applications of supercontinuum light source and all-fiber frequency comb are therefore possible based on such seed lasers. In addition, due to the bidirectional operation property of such devices, a bidirectional NPE mode-locked all-fiber laser is also possible. Very recently, by properly designing the coating of a 45° TFG, one could obtain controllable oscillation through $s$-light radiation. Such engineered spectral oscillation from air/cladding boundary has been proved to be useful as an optical sensor, with further optimization, it could replace the current Lyot filter scheme to implement a more compact multiwavelength or high repetition rate mode-locked fiber laser. We believe with nanomaterial integration/manipulation and laser micro-machining, the versatility and functionality of 45° TFG would be further improved, therefore more advanced mode-locked fiber lasers based on 45° TFG device will be possible in the future.

## ACKNOWLEDGEMENT

The authors would like to thank Dr. Kaiming Zhou and Prof. Lin Zhang from Aston Institute of Photonic Technologies at Aston University, Birmingham, UK for strong support. The authors would also like to acknowledge support from National Nature Science Foundation of China (61975107,62075071,62135007 and 61605107), and the "111" project (D20031), Natural Science Foundation of Shanghai (20ZR1471500) and the open fund of Key Laboratory of Space Active Opto-electronics Technology, Chinese Academy of Sciences (2021-ZDKF-1).

## REFERENCES

Bachim, B. L. and T. K. Gaylord. 2003. "Polarization-dependent loss and birefringence in long-period fiber gratings." *Applied Optics* 42 (34): 6816. doi:10.1364/AO.42.006816.

Bao, Q., H. Zhang, B. Wang, Z. Ni, C. H. Y. X. Lim, Y. Wang, D. Y. Tang, and K. P. Loh. 2011. "Broadband graphene polarizer." *Nature Photonics* 5 (7): 411–415. doi:10.1038/nphoton.2011.102.

Bernier, M., D. Faucher, R. Vallée, A. Saliminia, G. Androz, Y. Sheng, and S. L. Chin. 2007. "Bragg gratings photoinduced in ZBLAN fibers by femtosecond pulses at 800 Nm." *Optics Letters* 32 (5): 454–456. doi:10.1364/OL.32.000454.

Bernier, M., M. El-Amraoui, J. F. Couillard, Y. Messaddeq, and R. Vallée. 2012. "Writing of Bragg gratings through the polymer jacket of low-loss $As_2S_3$ fibers using femtosecond pulses at 800 Nm." *Optics Letters* 37 (18): 3900–3902. doi:10.1364/OL.37.003900.

Bewersdorf, J. and S. W. Hell. 1998. "Picosecond pulsed two-photon imaging with repetition rates of 200 and 400 MHz." *Journal of Microscopy* 191: 28–38. doi:10.1046/j.1365-2818.1998.00379.x.

Brasselet, S. 2011. "Polarization-resolved nonlinear microscopy: Application to structural molecular and biological imaging." *Advances in Optics and Photonics* 3 (3): 205. doi:10.1364/AOP.3.000205.

Cheng, H., W. Wang, Y. Zhou, T. Qiao, W. Lin, S. Xu, and Z. Yang. 2017. "5 GHz fundamental repetition rate, wavelength tunable, all-fiber passively mode-locked Yb-fiber laser." *Optics Express* 25 (22): 27646–27651. doi:10.1364/OE.25.027646.

Cheng, X., Q. Huang, C. Zou, C. Mou, Z. Yan, K. Zhou, and L. Zhang. 2019. "Pump-controlled flexible generation between dissipative soliton and noise-like pulses from a mode-locked Er-doped fiber laser." *Applied Optics* 58 (14): 3932. doi:10.1364/AO.58.003932.

Cheng, X., Q. Huang, Z. Huang, Q. Song, C. Zou, L. Zhao, C. Mou, Z. Yan, K. Zhou, and L. Zhang. 2020. "Multi-shuttle behavior between dissipative solitons and noise-like pulses in an all-fiber laser." *Journal of Lightwave Technology* 38 (8): 2471–2476. doi:10.1109/JLT.2020.2973309.

Chong, A., J. Buckley, W. Renninger, and F. Wise. 2006. "All-normal-dispersion femtosecond fiber laser." *Optics Express* 14 (21): 10095–10100. doi:10.1364/OE.14.010095.

Dennis, M. L. and I. N. Duling. 1994. "Experimental study of sideband generation in femtosecond fiber lasers." *IEEE Journal of Quantum Electronics* 30 (6): 1469–1477. doi:10.1109/3.299472.

Dobrowolski, J. A. and A. Waldorf. 1981. "High-performance thin film polarizer for the UV and visible spectral regions." *Applied Optics* 20 (1): 111–116. doi:10.1364/AO.20.000111.

Dong, X., P. Shum, N. Q. Ngo, H.-Y. Tam, and X. Dong. 2005. "Output power characteristics of tunable erbium-doped fiber ring lasers." *Journal of Lightwave Technology* 23 (3): 1334–1341. doi:10.1109/JLT.2004.839986.

Dong, T., J. Lin, G. Chun, P. Yao, and X. Lixin. 2021. "Noise-like square pulses in both normal and anomalous dispersion regimes." *IEEE Photonics Journal* 13 (2): 1–8. doi:10.1109/JPHOT.2021.3066792.

Doran, N. J. and D. Wood. 1988. "Nonlinear-optical loop mirror." *Optics Letters* 13 (1): 56–58. doi:10.1364/OL.13.000056.

Du, Y., X. Shu, and X. Zuowei. 2016. "All-fiber passively mode-locked laser based on a chiral fiber grating." *Optics Letters* 41 (2): 360–363. doi:10.1364/OL.41.000360.

Du, Y., M. Han, P. Cheng, and X. Shu. 2019. "Pulsating soliton with broadened Kelly sidebands in an ultrafast fiber laser." *Optics Letters* 44 (16): 4087–4090. doi:10.1364/OL.44.004087.

Eickhoff, W. 1980. "In-line fibre-optic polariser." *Electronics Letters* 16 (20): 762–764. doi:10.1049/el:19800541.

Enokihara, A., M. Izutsu, and T. Sueta. 1987. "Optical fiber sensors using the method of polarization-rotated reflection." *Journal of Lightwave Technology* 5 (11): 1584–1590. doi:10.1109/JLT.1987.1075449.

Fermann, M. E., F. Haberl, M. Hofer, and H. Hochreiter. 1990. "Nonlinear amplifying loop mirror." *Optics Letters* 15 (13): 752–754. doi:10.1364/OL.15.000752.

Fu, B., D. Popa, Z. Zhao, S. A. Hussain, E. Flahaut, T. Hasan, G. Soavi, and A. C. Ferrari. 2018. "Wavelength tunable soliton rains in a nanotube-mode locked Tm-doped fiber laser." *Applied Physics Letters* 113 (19): 193102. doi:10.1063/1.5047492.

Gao, X., Z. Zhao, Z. Zhao, Z. Cong, G. Gao, A. Zhang, H. Guo, G. Yao, and Z. Liu. 2021. "Stable 5-GHz fundamental repetition rate passively SESAM mode-locked Er-doped silica fiber lasers." *Optics Express* 29 (6): 9021–9029. doi:10.1364/OE.414779.

Gomes, L. A., L. Orsila, T. Jouhti, and O. G. Okhotnikov. 2004. "Picosecond SESAM-based Ytterbium mode-locked fiber lasers." *IEEE Journal of Selected Topics in Quantum Electronics* 10 (1): 129–136. doi:10.1109/JSTQE.2003.822918.

Grelu, P. and N. Akhmediev. 2012. "Dissipative solitons for mode-locked lasers." *Nature Photonics* 6 (2): 84–92. doi:10.1038/nphoton.2011.345.

Guesmi, K., Y. Meng, A. Niang, P. Mouchel, M. Salhi, F. Bahloul, R. Attia, and F. Sanchez. 2014. "1.6 Mm emission based on linear loss control in a Er:Yb doped double-clad fiber laser." *Optics Letters* 39 (22): 6383–6386. doi:10.1364/OL.39.006383.

Han, Y., and G. Li. 2005. "Coherent optical communication using polarization multiple-input-multiple-output." *Optics Express* 13 (19): 7527–7534. doi:10.1364/OPEX.13.007527.

Haus, H. A. and W. S. Wong. 1996. "Solitons in optical communications." *Reviews of Modern Physics* 68 (2): 423–444. doi:10.1103/RevModPhys.68.423.

He, X., Z.-b. Liu, and D. N. Wang. 2012. "Wavelength-tunable, passively mode-locked fiber laser based on graphene and chirped fiber Bragg grating." *Optics Letters* 37 (12): 2394–2396. doi:10.1364/OL.37.002394.

He, J., X. Baijie, X. Xizhen, C. Liao, and Y. Wang. 2021. "Review of femtosecond-laser-inscribed fiber Bragg gratings: Fabrication technologies and sensing applications." *Photonic Sensors* 11 (2): 203–226. doi:10.1007/s13320-021-0629-2.

Huang, Q., C. Jiang, C. Zou, Z. Huang, C. Mou, and Y. Liu. 2019a. "1.94 GHz Passively harmonic mode-locked all-fiber laser using polarization-maintaining helical long-period grating." In *Conference on Lasers and Electro-Optics*, JW2A.88. San Jose, California: OSA. doi:10.1364/CLEO_AT.2019.JW2A.88.

Huang, Q., C. Zou, C. Mou, X. Guo, Z. Yan, K. Zhou, and L. Zhang. 2019b. "23 MHz widely wavelength-tunable L-band dissipative soliton from an all-fiber Er-doped laser." *Optics Express* 27 (14): 20028. doi:10.1364/OE.27.020028.

Huang, Z., Q. Huang, A. Theodosiou, X. Cheng, C. Zou, L. Dai, K. Kalli, and C. Mou. 2019c. "All-fiber passively mode-locked ultrafast laser based on a femtosecond-laser-inscribed in-fiber brewster device." *Optics Letters* 44 (21): 5177. doi:10.1364/OL.44.005177.

Huang, Z., S. Boscolo, Q. Huang, Z. Xing, Z. Yan, T. Chen, Y. Liu, and C. Mou. 2021a. "Generation of 64-Fs L-band stretched pulses from an all-fibre Er-doped laser." *Optics Express* 29 (22): 34892. doi:10.1364/OE.434546.

Huang, Z., Q. Huang, A. Theodosiou, K. Kalli, S. Li, N. Chen, T. Chen, and C. Mou. 2021b. "Femtosecond laser direct inscribed 45° tilted fiber grating for a net-normal-dispersion mode-locked Er-doped fiber laser." *Optics & Laser Technology* 143 (November): 107358. doi:10.1016/j.optlastec.2021.107358.

Ioannou, A., A. Theodosiou, C. Caucheteur, and K. Kalli. 2017. "Direct writing of plane-by-plane tilted fiber Bragg gratings using a femtosecond laser." *Optics Letters* 42 (24): 5198–5201. doi:10.1364/OL.42.005198.

Jiang, C., Y. Liu, Y. Zhao, C. Mou, and T. Wang. 2019. "Helical long-period gratings inscribed in polarization-maintaining fibers by CO2 laser." *Journal of Lightwave Technology* 37 (3): 889–896. doi:10.1109/JLT.2018.2883376.

Kanagaraj, N., A. Theodosiou, J. Aubrecht, P. Peterka, M. Kamradek, K. Kalli, I. Kasik, and P. Honzatko. 2019. "All fiber mode-locked thulium-doped fiber laser using a novel femtosecond-laser-inscribed 45°-plane-by-plane-tilted fiber grating." *Laser Physics Letters* 16 (9): 095104. doi:10.1088/1612-202X/ab39db.

Kawagoe, H., S. Ishida, M. Aramaki, Y. Sakakibara, E. Omoda, H. Kataura, and N. Nishizawa. 2014. "Development of a high power supercontinuum source in the 1.7 Mm wavelength region for highly penetrative ultrahigh-resolution optical coherence tomography." *Biomedical Optics Express* 5 (3): 932–943. doi:10.1364/BOE.5.000932.

Kim, J. and Y. Song. 2016. "Ultralow-noise mode-locked fiber lasers and frequency combs: Principles, status, and applications." *Advances in Optics and Photonics* 8 (3): 465–540. doi:10.1364/AOP.8.000465.

Kopp, V. I., V. M. Churikov, J. Singer, N. Chao, D. Neugroschl, and A. Z. Genack. 2004. "Chiral fiber gratings." *Science* 305 (5680): 74–75. doi:10.1126/science.1097631.

Kurkov, A. S., M. Douay, O. Duhem, B. Leleu, J. F. Henninot, J. F. Bayon, and L. Rivoallan. 1997. "Long-period fibre grating as a wavelength selective polarisation element." *Electronics Letters* 33 (7): 616–617. doi:10.1049/el:19970422.

Lecourt, J.-B., C. Duterte, F. Narbonneau, D. Kinet, Y. Hernandez, and D. Giannone. 2012. "All-normal dispersion, all-fibered pm laser mode-locked by SESAM." *Optics Express* 20 (11): 11918–11923. doi:10.1364/OE.20.011918.

Li, S. and K. T. Chan. 1999. "Actively mode-locked erbium fiber ring laser using a Fabry–Perot semiconductor modulator as mode locker and tunable filter." *Applied Physics Letters* 74 (19): 2737–2739. doi:10.1063/1.123998.

Li, J., Z. Yan, Z. Sun, H. Luo, Y. He, Z. Li, Y. Liu, and L. Zhang. 2014. "Thulium-doped all-fiber mode-locked laser based on NPR and 45°-tilted fiber grating." *Optics Express* 22 (25): 31020. doi:10.1364/OE.22.031020.

Li, J., Y. Wang, H. Luo, Y. Liu, Z. Yan, Z. Sun, and L. Zhang. 2019. "Kelly sideband suppression and wavelength tuning of a conventional soliton in a Tm-doped hybrid mode-locked fiber laser with an all-fiber lyot filter." *Photonics Research* 7 (2): 103. doi:10.1364/PRJ.7.000103.

Li, L. and L. Xiao. 2019. "Plasmonic nodeless hollow-core photonic crystal fibers for in-fiber polarizers." *Journal of Lightwave Technology* 37 (20): 5199–5211. doi:10.1109/JLT.2019.2930075.

Lin, S.-F., and G.-R. Lin. 2014. "Dual-band wavelength tunable nonlinear polarization rotation mode-locked erbium-doped fiber lasers induced by birefringence variation and gain curvature alteration." *Optics Express* 22 (18): 22121–22132. doi:10.1364/OE.22.022121.

Lin, Y., Z. Huang, Q. Huang, L. Dai, Q. Song, Z. Yan, C. Mou, K. Zhou, and L. Zhang. 2020. "Pump-controlled wavelength switchable dissipative soliton mode-locked Yb-doped fiber laser using a 45° tilted fiber grating." *Optik* 222 (November): 165383. doi:10.1016/j.ijleo.2020.165383.

Lin, Y., Z. Huang, Q. Huang, L. Dai, Y. Bao, T. Chen, Q. Song, et al. 2021a. "(INVITED) Switchable multi-wavelength mode-locked Yb-doped fiber laser using a polarization maintaining 45°-tilted fiber gratings based lyot filter." *Results in Optics* 3 (May): 100071. doi:10.1016/j.rio.2021.100071.

Lin, Y., Z. Huang, Q. Huang, L. Dai, Z. Xing, Z. Yan, and C. Mou. 2021b. "Pulse state switchable ytterbium-doped fiber laser based on lyot filter." *Chinese Journal of Lasers* 48 (19): 1901004. doi:10.3788/CJL202148.1901004.

Ling, Y., Q. Huang, C. Zou, Z. Xing, Z. Yan, C. Zhao, K. Zhou, L. Zhang, and C. Mou. 2019. "L-band GHz femtosecond passively harmonic mode-locked Er-doped fiber laser based on nonlinear polarization rotation." *IEEE Photonics Journal* 11 (4): 1–7. doi:10.1109/JPHOT.2019.2927771.

Ling, Y., Q. Huang, Q. Song, Z. Yan, C. Mou, K. Zhou, and L. Zhang. 2020. "Intracavity birefringence-controlled GHz-tuning range passively harmonic mode-locked fiber laser based on NPR." *Applied Optics* 59 (22): 6724. doi:10.1364/AO.398960.

Liu, X., H. Wang, Z. Yan, Y. Wang, W. Zhao, W. Zhang, L. Zhang, et al. 2012. "All-fiber normal-dispersion single-polarization passively mode-locked laser based on a 45°-tilted fiber grating." *Optics Express* 20 (17): 19000. doi:10.1364/OE.20.019000.

Liu, X., H. Wang, Y. Wang, W. Zhao, W. Zhang, X. Tan, Z. Yang, et al. 2013a. "Bound dissipative-pulse evolution in the all-normal dispersion fiber laser using a 45° tilted fiber grating." *Laser Physics Letters* 10 (9): 095103. doi:10.1088/1612-2011/10/9/095103.

Liu, X., D. Han, Z. Sun, C. Zeng, H. Lu, M. Dong, Y. Cui, and F. Wang. 2013b. "Versatile multi-wavelength ultrafast fiber laser mode-locked by carbon nanotubes." *Scientific Reports* 3 (1): 2718. doi:10.1038/srep02718.

Liu, X., H. Wang, Y. Wang, Z. Yan, and L. Zhang. 2015. "Single-polarization, dual-wavelength mode-locked Yb-doped fiber laser by a 45°-tilted fiber grating." *Laser Physics Letters* 12 (6): 065102. doi:10.1088/1612-2011/12/6/065102.

Lopez, A. G. and H. G. Craighead. 1998. "Wave-plate polarizing beam splitter based on a form-birefringent multilayer grating." *Optics Letters* 23 (20): 1627–1629. doi:10.1364/OL.23.001627.

Lu, B., C. Zou, Q. Huang, Z. Yan, Z. Xing, M. Al Araimi, A. Rozhin, K. Zhou, L. Zhang, and C. Mou. 2019. "Widely wavelength-tunable mode-locked fiber laser based on a 45°-tilted fiber grating and polarization maintaining fiber." *Journal of Lightwave Technology* 37 (14): 3571–3578. doi:10.1109/JLT.2019.2918016.

Luo, Z.-C., A.-P. Luo, W.-C. Xu, H.-S. Yin, J.-R. Liu, Q. Ye, and Z.-J. Fang. 2010. "Tunable multiwavelength passively mode-locked fiber ring laser using intracavity birefringence-induced comb filter." *IEEE Photonics Journal* 2 (4): 571–577. doi:10.1109/JPHOT.2010.2051023.

Morin, F., F. Druon, M. Hanna, and P. Georges. 2009. "Microjoule femtosecond fiber laser at 1.6 μm for corneal surgery applications." *Optics Letters* 34 (13): 1991–1993. doi:10.1364/OL.34.001991.

Mou, C., K. Zhou, L. Zhang, and I. Bennion. 2009. "Characterization of 45°-tilted fiber grating and its polarization function in fiber ring laser." *Journal of the Optical Society of America B* 26 (10): 1905. doi:10.1364/JOSAB.26.001905.

Mou, C., H. Wang, B. G. Bale, K. Zhou, L. Zhang, and I. Bennion. 2010. "All-fiber passively mode-locked femtosecond laser using a 45°-tilted fiber grating polarization element." *Optics Express* 18 (18): 18906. doi:10.1364/OE.18.018906.

Ortega, B., L. Dong, W. F. Liu, J. P. de Sandro, L. Reekie, S. I. Tsypina, V. N. Bagratashvili, and R. I. Laming. 1997. "High-performance optical fiber polarizers based on long-period gratings in birefringent optical fibers." *IEEE Photonics Technology Letters* 9 (10): 1370–1372. doi:10.1109/68.623266.

Pei, C., L. Yang, G. Wang, Y. Wang, X. Jiang, Y. Hao, Y. Li, and J. Yang. 2015. "Broadband graphene/glass hybrid waveguide polarizer." *IEEE Photonics Technology Letters* 27 (9): 927–930. doi:10.1109/LPT.2015.2398452.

Peng, G. D., T. Tjugiarto, and P. L. Chu. 1990. "Polarisation beam splitting using twin-elliptic-core optical fibres." *Electronics Letters* 26 (10): 682–683. doi:10.1049/el:19900446.

Peng, J., M. Sorokina, S. Sugavanam, N. Tarasov, D. V. Churkin, S. K. Turitsyn, and H. Zeng. 2018. "Real-time observation of dissipative soliton formation in nonlinear polarization rotation mode-locked fibre lasers." *Communications Physics* 1 (1): 1–8. doi:10.1038/s42005-018-0022-7.

Posner, M. T., N. Podoliak, D. H. Smith, P. L. Mennea, P. Horak, C. B. E. Gawith, P. G. R. Smith, and J. C. Gates. 2019. "Integrated polarizer based on 45° tilted gratings." *Optics Express* 27 (8): 11174. doi:10.1364/OE.27.011174.

Qin, H., Q. He, Z. Xing, X. Guo, Z. Yan, Q. Sun, K. Zhou, H. Wang, D. Liu, and L. Zhang. 2019. "Numerical and experimental characterization of radiation mode of 45° tilted fiber grating." *Journal of Lightwave Technology*, 1–1. doi:10.1109/JLT.2019.2920680.

Schlager, J. B., P. D. Hale, and D. L. Franzen. 1993. "High-sensitivity optical sampling using an erbium-doped fiber laser strobe." *Microwave and Optical Technology Letters* 6 (15): 835–837. doi:10.1002/mop.4650061503.

Simpson, J., R. Stolen, F. Sears, W. Pleibel, J. MacChesney, and R. Howard. 1983. "A single-polarization fiber." *Journal of Lightwave Technology* 1 (2): 370–374. doi:10.1109/JLT.1983.1072129.

Srivastava, A. K., S. Radic, C. Wolf, J. C. Centanni, J. W. Sulhoff, K. Kantor, and Y. Sun. 2000. "Ultradense WDM transmission in L-band." *IEEE Photonics Technology Letters* 12 (11): 1570–1572. doi:10.1109/68.887758.

Suzuki, M., R. A. Ganeev, S. Yoneya, and H. Kuroda. 2015. "Generation of broadband noise-like pulse from Yb-doped fiber laser ring cavity." *Optics Letters* 40 (5). Optica Publishing Group: 804–807. doi:10.1364/OL.40.000804.

Tamura, K., H. A. Haus, and E. P. Ippen. 1992. "Self-starting additive pulse mode-locked erbium fibre ring laser." *Electronics Letters* 24 (28): 2226–2228. doi:10.1049/el:19921430.

Theodosiou, A., A. Lacraz, M. Polis, K. Kalli, M. Tsangari, A. Stassis, and M. Komodromos. 2016. "Modified Fs-laser inscribed FBG array for rapid mode shape capture of free-free vibrating beams." *IEEE Photonics Technology Letters* 28 (14): 1509–1512. doi:10.1109/LPT.2016.2555852.

Villarini, G. and W. F. Krajewski. 2010. "Review of the different sources of uncertainty in single polarization radar-based estimates of rainfall." *Surveys in Geophysics* 31 (1): 107–129. doi:10.1007/s10712-009-9079-x.

Vogler, N., S. Heuke, T. W. Bocklitz, M. Schmitt, and J. Popp. 2015. "Multimodal imaging spectroscopy of tissue." *Annual Review of Analytical Chemistry* 8 (1): 359–387. doi:10.1146/annurev-anchem-071114-040352.

Wagener, J. L., P. F. Wysocki, M. J. F. Digonnet, H. J. Shaw, and D. J. DiGiovanni. 1993. "Effects of concentration and clusters in erbium-doped fiber lasers." *Optics Letters* 18 (23): 2014–2016. doi:10.1364/OL.18.002014.

Wang, J., A. P. Zhang, Y. H. Shen, H.-y. Tam, and P. K. A. Wai. 2015. "Widely tunable mode-locked fiber laser using carbon nanotube and LPG W-shaped filter." *Optics Letters* 40 (18): 4329–4332. doi:10.1364/OL.40.004329.

Wang, T., Z. Yan, Q. Huang, C. Zou, C. Mou, K. Zhou, and L. Zhang. 2018a. "Sub-70fs generation from passively mode locked erbium doped fiber laser using 45° tilted fiber grating." In *CLEO Pacific Rim Conference*, Th2A.3. Hong Kong: OSA. doi:10.1364/CLEOPR.2018.Th2A.3.

Wang, T., Z. Yan, Q. Huang, C. Zou, C. Mou, K. Zhou, and L. Zhang. 2018b. "Mode-locked erbium-doped fiber lasers using 45° tilted fiber grating." *IEEE Journal of Selected Topics in Quantum Electronics* 24 (3): 1–6. doi:10.1109/JSTQE.2017.2779038.

Wang, X., J. Lin, W. Sun, Z. Tan, R. Liu, and Z. Wang. 2020. "Polarization selectivity of the thin-metal-film plasmon-assisted fiber-optic polarizer." *ACS Applied Materials & Interfaces* 12 (28): 32189–32196. doi:10.1021/acsami.0c08274.

Williams, R. J., R. G. Krämer, S. Nolte, and M. J. Withford. 2013. "Femtosecond direct-writing of low-loss fiber Bragg gratings using a continuous core-scanning technique." *Optics Letters* 38 (11): 1918–1920. doi:10.1364/OL.38.001918.

Wu, C. and N. K. Dutta. 2000. "High-repetition-rate optical pulse generation using a rational harmonic mode-locked fiber laser." *IEEE Journal of Quantum Electronics* 36 (2): 145–150. doi:10.1109/3.823458.

Wu, X., D. Y. Tang, L. M. Zhao, and H. Zhang. 2010. "Mode-locking of fiber lasers induced by residual polarization dependent loss of cavity components." *Laser Physics* 20 (10): 1913–1917. doi:10.1134/S1054660X10190187.

Xing, L., W. Zou, G. Yang, and J. Chen. 2015. "Direct generation of 148 Nm and 44.6 Fs pulses in an erbium-doped fiber laser." *IEEE Photonics Technology Letters* 27 (1): 93–96. doi:10.1109/LPT.2014.2362514.

Xing, Z., Z. Yan, Q. Song, C. Mou, Q. Sun, D. Liu, and L. Zhang. 2020. "All-fiber mode-locked laser utilizing 45°-tilted fiber grating-based polarization beam splitter." *IEEE Photonics Technology Letters* 32 (21): 1389–1392. doi:10.1109/LPT.2020.3024802.

Xue, L., B. S. M. Timoteo, W. Qiu, and Z. Wang. 2022. "Broadband circular polarizer based on chirped double-helix chiral fiber grating." *Materials* 15 (9): 3366. doi:10.3390/ma15093366.

Yamada, I., K. Takano, M. Hangyo, M. Saito, and W. Watanabe. 2009. "Terahertz wire-grid polarizers with micrometer-pitch Al gratings." *Optics Letters* 34 (3): 274–276. doi:10.1364/OL.34.000274.

Yan, Z., C. Mou, K. Zhou, X. Chen, and L. Zhang. 2011. "UV-inscription, polarization-dependant loss characteristics and applications of 45$^{\circ}$ tilted fiber gratings." *Journal of Lightwave Technology* 29 (18): 2715–2724. doi:10.1109/JLT.2011.2163196.

Yan, Z., C. Mou, H. Wang, K. Zhou, Y. Wang, W. Zhao, and L. Zhang. 2012. "All-fiber polarization interference filters based on 45°-tilted fiber gratings." *Optics Letters* 37 (3): 353. doi:10.1364/OL.37.000353.

Yan, Z., C. Mou, Y. Wang, J. Li, Z. Zhang, X. Liu, K. Zhou, and L. Zhang. 2016. "45°-Tilted fiber gratings and their application in ultrafast fiber lasers." In *Fiber Laser*, M. C. Paul. (ed.) InTech. doi:10.5772/61739.

Zhang, J., X. Qiao, F. Liu, Y. Weng, R. Wang, Y. Ma, Q. Rong, M. Hu, and Z. Feng. 2011. "A tunable erbium-doped fiber laser based on an MZ interferometer and a birefringence fiber filter." *Journal of Optics* 14 (1): 015402. doi:10.1088/2040-8978/14/1/015402.

Zhang, Z., C. Mou, Z. Yan, K. Zhou, L. Zhang, and S. Turitsyn. 2013. "Sub-100 Fs mode-locked erbium-doped fiber laser using a 45°-tilted fiber grating." *Optics Express* 21 (23): 28297. doi:10.1364/OE.21.028297.

Zhang, Z., J. Gan, T. Yang, Y. Wu, Q. Li, S. Xu, and Z. Yang. 2015a. "All-fiber mode-locked laser based on microfiber polarizer." *Optics Letters* 40 (5): 784–787. doi:10.1364/OL.40.000784.

Zhang, Z., Z. Yan, K. Zhou, and L. Zhang. 2015b. "All-fiber 250 MHz fundamental repetition rate pulsed laser with tilted fiber grating polarizer." *Laser Physics Letters* 12 (4): 045102. doi:10.1088/1612-2011/12/4/045102.

Zhao, L. M., D. Y. Tang, H. Zhang, and X. Wu. 2008. "Polarization rotation locking of vector solitons in a fiber ring laser." *Optics Express* 16 (14): 10053–10058. doi:10.1364/OE.16.010053.

Zhao, X., Z. Zheng, L. Liu, Y. Liu, Y. Jiang, X. Yang, and J. Zhu. 2011. "Switchable, dual-wavelength passively mode-locked ultrafast fiber laser based on a single-wall carbon nanotube modelocker and intracavity loss tuning." *Optics Express* 19 (2): 1168–1173. doi:10.1364/OE.19.001168.

Zhao, W., Q. Huang, K. Li, C. Gao, X. Cheng, Y. Yan, Q. Guo, X. Sun, and C. Mou. 2021. "High-energy noise-like pulses generated by an erbium-doped fiber laser incorporating a PbS quantum-dot polystyrene composite film." *Journal of Physics: Photonics* 3 (2): 024015. doi:10.1088/2515-7647/abedd1.

Zhou, K., G. Simpson, X. Chen, L. Zhang, and I. Bennion. 2005. "High extinction ratio in-fiber polarizers based on 45° tilted fiber Bragg gratings." *Optics Letters* 30 (11): 1285. doi:10.1364/OL.30.001285.

Zhou, X., M. Qiu, Y. Qian, M. Chen, Z. Zhang, and L. Zhang. 2021. "Microfiber-based polar-ization beam splitter and its application for passively mode-locked all-fiber laser." *IEEE Journal of Selected Topics in Quantum Electronics* 27 (2): 1–6. doi:10.1109/JSTQE.2020.3012667.

Zou, C., T. Wang, Z. Yan, Q. Huang, M. AlAraimi, A. Rozhin, and C. Mou. 2018. "Wavelength-tunable passively mode-locked erbium-doped fiber laser based on carbon nanotube and a 45° tilted fiber grating." *Optics Communications* 406 (January): 151–157. doi:10.1016/j.optcom.2017.06.006.

Zou, M., Y. Ran, J. Hu, Z. Xing, Z. Yan, C. Liu, Q. Sun, and D. Liu. 2020. "Multiwavelength mode-locked fiber laser based on an all fiber lyot filter." *IEEE Photonics Technology Letters* 32 (22): 1419–1422. doi:10.1109/LPT.2020.3029089.

# 3 Polarization and Color Domains in Fiber Lasers

Yichang Meng, Ahmed Nady, Georges Semaan,
A. Komarov, M. Kemel, M. Salhi and F. Sanchez

## CONTENTS

3.1 Introduction ...................................................................................................... 103
3.2 Polarization Domain Walls in Fiber Lasers ...................................................... 104
3.3 Color Domain Walls in Fiber Lasers ................................................................ 109
3.4 Conclusions ...................................................................................................... 119
References ................................................................................................................ 119

## 3.1 INTRODUCTION

Polarization domains (PD) are domains with different polarization eigenstates that are separated spatially or temporally. In 1987, Zakharov pioneered the concept in nonlinear optics by investigating the interaction of two counterpropagating beams in a nonlinear Kerr medium [1]. He demonstrated that cross-polarization modulation resulted in the development of localized polarization domains corresponding to polarization states, thereby minimizing the Hamiltonian of the system. This opened the way for many investigations on the vectorial characteristics of optical waves propagating in nonlinear and conservative media. It started with vector solitons, more specifically bound vector solitary waves [2–8]. Wabnitz established the concept of polarization domain walls (PDW) in 1990 [9] by studying the dynamics of two counterpropagating waves along a nonlinear anisotropic fiber. According to theory, if the self-induced polarization change (SPM) is ignored, the asymptotic evolution of two counterpropagating waves is a distribution of domains with constant polarization along one of the medium's eigenpolarization states. These domains are separated by walls. After then, the case of an isotropic fiber with optical Kerr nonlinearity was considered [10]. Haelterman first proposed the idea of a polarization domain wall soliton [11]. It was discovered that vector dark solitary waves in Kerr media are related to polarization domain walls and that PDW is a type of vector dark soliton. Domain walls are not always related to polarization eigenstates. Indeed, the possibility to generate frequency domain walls has been theoretically predicted from two coupled nonlinear Schrödinger equations (NLSE) [12]. Pitois reported the first experimental demonstration of PDWs in optical fibers [13]. He investigated two counterpropagating circularly polarized beams in a single-mode optical fiber. When the polarizations are corotating, a stable region of uniform polarization is observed over the fiber length. Meanwhile, for counterrotating beams, polarization switching

DOI: 10.1201/9781003206767-3

**103**

happens because of an unstable arrangement. As a result of this switching, a stationary PDW emerges, connecting two counterrotating polarizations. The experimental results are consistent with a model based on two NLSEs with negligible SPM and no Group Velocity Dispersion (GVD). The polarization switching that occurs when distinct input polarizations are launched at both fibers' ends was considered in a generalization of this study [14]. The switches exist because corotating circular polarization is the only stable state. As a result, the system changes to a stable state if the initial polarizations do not corotate. Finally, between the two separate polarization states, PDW occurs. This process manifests itself as a form of polarization attraction toward the stable polarization state, as elegantly demonstrated in reference [15]. In fact, the authors considered a fiber with elliptical birefringence, optical Kerr nonlinearity (SPM and XPM, cross-phase modulation), and no GVD. They investigated the coupling of two copropagating channels with distinct optical wavelengths, and theoretically demonstrated the existence of domain wall solitons, which represent the locked temporal switching of both beams' polarization states. They also used an unpolarized probe beam coupled to a polarized pump beam to demonstrate the phenomenon of polarization attraction. In this situation, the probe beam's polarization switches to that of the pump beam, giving rise to the concept of a lossless polarizer. Recently, PDW has been employed as data transmission polarization bits [16]. The authors consider the propagation of polarization domain walls containing binary encoded information over standard single-mode fiber. The injected signal is built as antiphase orthogonally polarized waves. It is demonstrated that when only one polarization is propagated in a 10 km fiber, the signal is rapidly degraded into a complex periodic pattern. In contrast, when both polarizations are injected, PDW propagate as a unit and do not suffer from any degradation. This is a remarkable outcome since it displays the resilience of PDWs. It's worth noting that while the original twin orthogonally polarized components are uncorrelated, they are strongly correlated at the fiber's output, implying that the initial signal converts into PDWs. This spontaneous emergence of polarization domains is referred to as polarization segregation. All prior results were obtained using conservative optical systems. PDW can be investigated in nonlinear optical passive cavities [17], which are not conservative because permanent external driving is required to compensate for cavity losses. PDW in fiber Kerr resonators will be discussed in detail in Chapter 4.

This chapter aims to present the recent developments in the field of domain walls in fiber lasers. We first consider the case of polarization domain walls followed by the case of color domains. The hybrid case of polarization-color-domain is also presented.

## 3.2 POLARIZATION DOMAIN WALLS IN FIBER LASERS

Fiber lasers are excellent tools for investigating polarization dynamics that often manifest as an antiphase dynamic between the two orthogonally polarized eigenmodes [18–21]. The physical mechanisms responsible for the coupling vary depending on the timescale over which the dynamic occurs. When the timescale of the antiphase dynamics relates to the gain dynamics, polarization switching is caused by cross-gain saturation between the two polarization eigenstates [18–20]. The cross-saturation of the gain is also responsible for the antiphase dynamics between two frequency super-modes in erbium-doped fiber lasers [22–24]. Roy was the first to experimentally report on the spontaneous fast polarization switching in a fiber laser with a round-trip period and to develop a theoretical model based on delay differential equations [21]. Essentially, this is analogous to polarization domains with antiphase dynamics between the two polarization components. However, in this case, the coupling between the two polarization components is due to cross-gain saturation through the amplifying medium rather than the optical Kerr effect. The observation of vector solitons in EDFL (Erbium-Doped Fiber Laser) was reported experimentally in 2008 [25]. The cavity is quasi-isotropic in that, despite having birefringence, it lacks a polarizing element, resulting in no polarization-dependent losses. A numerical solution is found for an adapted theoretical model based on two coupled nonlinear equations. The cross-phase modulation associated with the optical Kerr effect is shown to produce vector solitons while cross-saturation through the gain medium is not considered.

Although dark-dark pulses had previously been observed [26], the first experimental observation of PDW solitons was published in reference [27]. The authors employed an optical cavity similar to the one used in reference [21] and is schematically depicted in Figure 3.1. It consists of a unidirectional ring cavity with no dichroic element. The linear birefringence of the cavity was adjusted with a polarization controller. Square-pulse emission with antiphase phenomena has been detected between both linear polarizations. There is no evidence of such behavior on the overall intensity at low pump power, whereas a dip separates the two stable linear polarization states at high pump power (see Figure 3.2). The wavelengths of the two polarization states are different, implying that the coupling is incoherent [26, 27]. In addition, both cavity birefringence and pumping power influence the width and depth of the dip. As predicted in reference [11], the authors interpret the dip as a vector dark PDW soliton. They use a theoretical model [28] which confirms that a vector dark soliton could separate two stable linear polarizations.

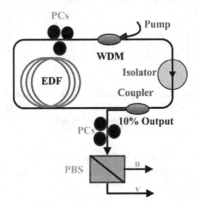

**FIGURE 3.1**  Typical experimental arrangement for PDW solitons in fiber lasers (from reference [26]). PCs are polarization controllers, and a polarization beam-splitter PBS allows the separation of both polarization components u and v.

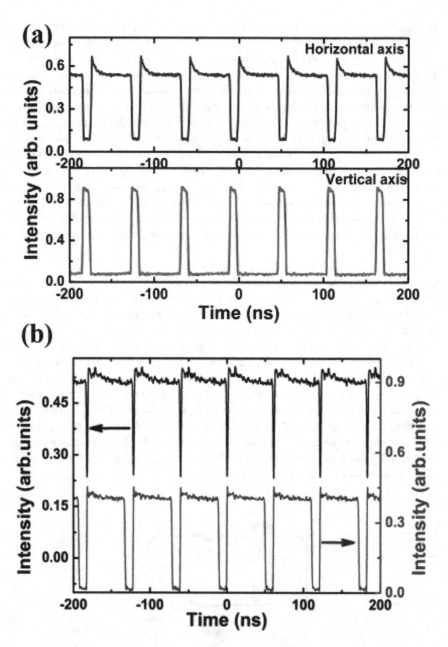

**FIGURE 3.2**   Polarization domain wall soliton emission of the laser. (a) Polarization resolved traces and (b) Total intensity (upper trace) and one of the polarized eigenmode (lower trace). (from reference [27].)

A simple theoretical model for polarization switching and antiphase dynamics at a temporal scale less than round-trip time of a fiber laser cavity is provided in reference [29]. The cavity has a low-level polarization-dependent loss and a significant birefringence. The GVD, Polarization Mode Dispersion (PMD), SPM, and XPM are all considered in the minimal model. Because the coherence term vanishes on average when the beat length is less than the cavity length, no coherent FWM term is considered. The formalism utilized is similar to the one proposed in reference [11], in that it is based on a mechanical approach with a derivation of a potential. Depending on the positions of the maxima, the separatrices linking them can be vector dark solitons or polarization domain walls. Experimental demonstration of black/bright vector solitons is pointed out in an EDFL close to PDW. Furthermore, starting from the Amplified Spontaneous Emission (ASE) noise, the authors develop a fully numerical laser model based on rate equations and two counterpropagating waves. This approach allowed for the first time to theoretically demonstrate the spontaneous formation of PDW transitions in a time scale far shorter than the round-trip time. It is also the first demonstration of PDW using the laser equations. The same group investigated the build-up of polarization domain walls and other associated antiphase polarization dynamics in a second study [30]. They considered a fiber laser with cross-polarization coupling and no polarization-dependent loss. Under the mean-field approximation, the authors propose a simple model with linear gain, spectral gain filtering, GVD, SPM, and XPM. Gain saturation and cross-gain saturation are included in the form of third-order terms. As a result, the final model is made up of two coupled Complex Ginzburg-Landau Equations (CGLEs). It is remarkable and surprising that a model based on the mean-field approximation may describe the phenomena on a time scale shorter than the round-trip time of the cavity. Numerical simulations predict the onset of chaotic dynamics, antiphase chaotic dynamics, and antiphase polarization domain walls. All these behaviors are achieved by simply increasing the strength of the polarization coupling. Antiphase periodic soliton trains can be generated along the polarization eigenaxes. According to theoretical simulations, polarization cross-coupling is required for the realization of the synchronization mechanism that results in antiphase dynamics. Experimental analysis is carried out employing an erbium-doped ring fiber laser. It is demonstrated that changing the orientation of an intracavity phase plate or the pumping power can cause a transition from disorder to antiphase pulse trains for specific cavity parameters, which is in good agreement with theoretical predictions. Other orientation of the waveplates produces different temporal patterns, such as vector dark/bright solitary waves, PDW, or antiphase synchronized chaotic evolution. The numerical calculations and experimental results are qualitatively consistent.

Additional physical insights have been reported in reference [31] due to the implementation of a fast detection system. The authors consider a quasi-isotropic anomalous dispersion cavity similar to that studied in reference [26]. At low pumping powers, the laser produces two distinct CW signals matching to the polarization eigenmodes. Above a certain threshold, abrupt switches occur between the two polarized modes, forming two polarization domains. By adjusting the linear cavity birefringence, the width of the domains can be adjusted. PDW splitting is also observed when the pumping power is increased. A detailed examination revealed that, depending on the pumping rate, a hole (black pulse) in the overall intensity can appear at one polarization-switching position. At very high pumping, scalar modulation instability arises, manifesting as scalar solitons polarized along the two eigenaxes, as seen in Figure 3.3. It is worth noting that PDW subsists in the presence of modulational instability. Previous results were achieved via incoherent coupling, i.e., the two polarization eigenmodes have different central wavelengths. Resonant coupling, on the other hand, allows the two polarization eigenmodes to be in-phase or antiphase while the polarization domains remain intact. A theoretical model based on coupled equations subjected to boundary conditions was developed. It includes GVD, SPM, XPM, birefringence, and linear gain while gain saturation and cross-gain saturation are not considered. Using the mean-field approximation, which involves averaging the field over one cavity round-trip, the model is reduced to two coupled NLSEs valid for a conservative system. This is possible because the gain is expected to perfectly compensate for the losses. Following the procedure proposed in reference [32], the stationary solutions verify two coupled equations for which a Hamiltonian can be derived, and by minimizing the latter, polarization domains are found. The resulting solutions allow an explanation of the key findings of the experiment: PDW generation, PDW splitting, and the presence of a dark pulse on the total intensity.

PDW has been observed using other rare-earth-doped fibers. A Yb-doped fiber laser without a polarization-dependent loss element is considered in reference [33]. Consequently, the laser emits radiations along the two polarization eigenaxes. Under specific conditions, the polarization-resolved signals exhibit antiphase dynamics with the emergence of polarization domains separated by dark pulses on the total

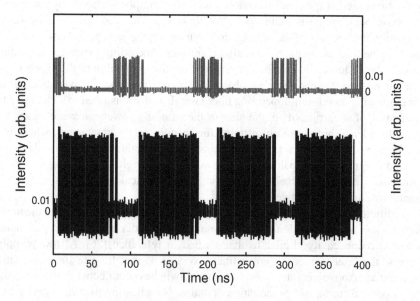

**FIGURE 3.3**    Polarization-resolved emission of the fiber laser points out that scalar solitons are formed inside the polarization domains. (from Tang et al. (2014) [31].)

intensity. This is analogous to the results obtained in the EDFL except that the optical spectrum is double-lobed. It is worth noting that the spectra associated with the two polarizations have different weighted spectral components. This is comparable to polarization-color domains discussed in the following section. Polarization-locked square pulses have been also reported.

PDW can be also obtained in fiber lasers using a two-dimensional-saturable absorber, as stated in reference [34]. The authors consider a classical EDFL without any polarization-dependent loss, but they include a real saturable absorber (SA). The SA is a topological insulator deposited on a microfiber. The laser can sustain two orthogonally polarized eigenmodes since the cavity is quasi-isotropic, allowing for polarization-resolved investigations. At low pumping powers, the laser emits incoherently coupled PDW for specified polarization controller orientations; however, at higher pumping powers, mode-locking occurs, yielding either polarization-locked vector dark pulses or polarization-locked noise-like pulses. The exact role of the SA on the PDW formation is not clearly identified since similar PDW can be obtained without SA, as previously reported in references [26, 29, 31] using a similar laser cavity. At higher pumping powers, the SA is responsible for the distinct vector-locked pulses; hence, it is expected that it does not influence the PDW formation.

Despite the fact that PDWs are stable solitary waves, they can break and split, eventually resulting in harmonic mode-locking of domain walls [35, 36]. In reference [35] the authors consider an EDFL in a bidirectional cavity without any polarizing element. The laser is passively mode-locked with a piece of Thulium-doped fiber acting as a saturable absorber. In contrast to previous works, the authors obtained in-phase dynamics between the two orthogonally polarized domains for both the clockwise and counterclockwise waves. As a result, harmonic mode-locking of PDW solitons is demonstrated. The features of polarization domain formation, domain splitting and shaping in dark-bright vector solitons have also been studied in a classical EDFL ring cavity without a polarization-dependent loss element in reference [36]. The authors investigate the influence of the average GVD on the polarization domain walls. The generation of PDW and dark-bright vector solitons is demonstrated to be independent of the cavity dispersion sign and to require weak cross-polarization coupling. The incoherently coupled dark-bright vector solitons are formed as a result of PD splitting. These dark-bright vector solitons can be thought of as the smallest stable domains in the cavity.

Before moving on to color domains, let us refer to reference [37] for a thorough examination of optical solitons in fiber lasers, including PDW.

## 3.3 COLOR DOMAIN WALLS IN FIBER LASERS

As mentioned in the introduction, domain walls in fiber lasers are not always associated with two polarization eigenmodes. In the context of cross-gain saturation, the antiphase dynamic between two wavelengths has been intensively examined in the EDFL [22–24]. Such dynamics occur on a time scale far longer than the round-trip time. By analogy with the polarization dynamics that occur on different time scales, under appropriate conditions, the coupling between the two wavelengths can be expected to lead to the emergence of wavelength domain walls. Dual-wavelength domain wall solitons were first reported in 2011 in dichroic ring laser cavity containing a polarizer [38]. Such configuration does not favor vector solitons or polarization domain walls. Due to the significant birefringence of the cavity, birefringent spectral filtering occurs and can be exploited to force the laser to operate in a dual-wavelength mode. Dark pulse emission is obtained on the total intensity. As shown in Figure 3.4, spectral analysis reveals that laser output alternates between two wavelengths, with the black pulse occurring at the point where the laser shifts from one wavelength to the other. The dark pulses are thought to be dual-wavelength domain wall solitons, according to the authors. The switching is not attributed to the gain competition but to the coupling through the optical Kerr nonlinearity. A numerical solution is found for an adapted theoretical model based on two coupled CGLEs. The model considers the birefringence, GVD, Third Order Dispersion (TOD), SPM, XPM, gain, and spectral gain filtering. Because no cross-gain saturation is considered, the coupling between the two wavelengths appears essentially in the XPM term. With appropriate initial conditions, it is theoretically demonstrated that domain wall soliton occurs and manifests as a dark pulse separating the two wavelengths. Similar experimental results were obtained in reference [39] in which the authors showed that by adjusting the polarization controllers, the duty cycle of the wavelength resolved signals could be adjusted.

Dual-wavelength domain wall solitons have been also reported in a Tm-doped fiber laser in a configuration, including a polarization-dependent isolator and a piece of HNLF (High Nonlinear Fiber) [40]. The latter induces large nonlinear effects whereas the former provides strong polarization-dependent losses (PDL). In principle, vector solitary waves cannot exist in a dichroic cavity. However, the authors report PDW, although each polarization has its own frequency as a result of a Loyt filter effect caused by a small piece of PMF (polarization-maintaining fiber with low beating length). Under such conditions, the domain walls are closer to wavelength domain walls rather than to polarization domain walls. One can expect that the formation of such domain walls is mainly due to cross-gain saturation between the two spectral components [23, 24] rather than polarization mode coupling.

In 2018, the concept of color domains (CD) in fiber lasers was introduced for the first time [41].

It consists of wavelength domains involving a condensate phase of solitons. In contrast to PDW, which requires a very low polarization-dependent loss, the optical cavity generating color domains includes a polarization-sensitive isolator as it is schematically depicted in Figure 3.5. Because long cavities support multiple-wavelength operation and narrow spectral bandwidth, which is required for CD operation, the overall length of the cavity has been set at 296 m. Indeed, the resulting spectral filtering effect based on the corresponding Lyot filter is adapted for CD generation and the PC aids in fine-tuning the spectral characteristics of the filter.

Mode-locking is achieved with a suitable orientation of the polarization controllers and a specific pumping power, resulting in a large, condensed soliton phase referred to as single color domain (SCD). It has typically one central wavelength and the temporal profile exhibits a square envelope containing many solitons, as evidenced by the autocorrelation trace. At a fixed orientation of the PC, the width of the SCD increases while the spectral bandwidth remains stable, as shown in Figure 3.6a.

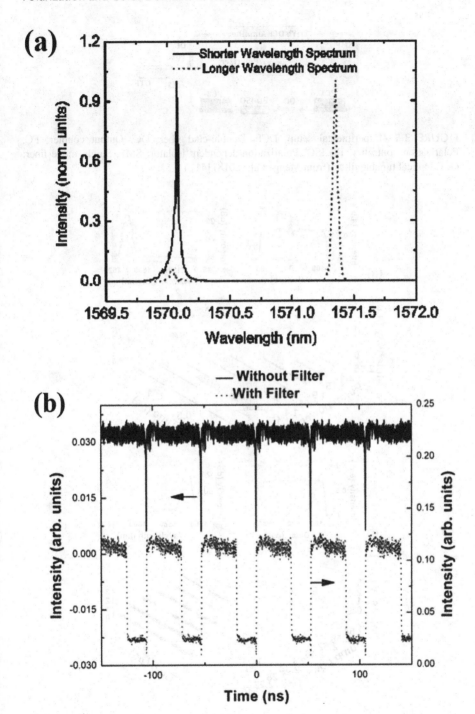

**FIGURE 3.4** (a) Wavelength resolved spectra and (b) Oscilloscope trace of the total (upper trace) and one wavelength laser emission (lower trace). (from Zhang et al. (2011) [38].)

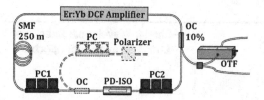

**FIGURE 3.5** Experimental setup. DCF, Double-clad fiber; OC, Output coupler; PC, Polarization controller; PD-ISO, Polarization-dependent isolator; SMF, Single mode fiber; OTF, Optical tunable filter. (from Meng et al. (2018) [41].)

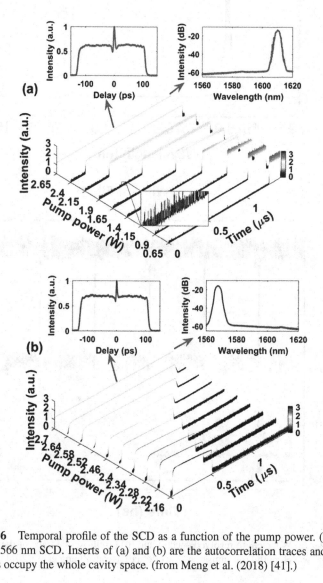

**FIGURE 3.6** Temporal profile of the SCD as a function of the pump power. (a) 1,610 nm SCD. (b) 1,566 nm SCD. Inserts of (a) and (b) are the autocorrelation traces and the spectra when SCDs occupy the whole cavity space. (from Meng et al. (2018) [41].)

The SCD operation is a dissipative regime in which a square envelope of solitons and large background coexist. This derives from a weak mode-locking due to the narrow-band filtering and peak power clamping effect, while the background is formed by the radiation of solitons in motion and other CW components [42, 43]. As illustrated in Figure 3.6a, when the pump power is high enough, the SCD stretches to occupy all the cavity space and the duty cycle becomes equal to 1. When the orientation of PCs is altered, SCD operation at 1,566 nm can also be observed, as shown in Figure 3.6b. The SCDs at 1,566 nm have roughly the same characteristics as the SCDs at 1,610 nm. By adjusting PC2, the peak power and the duty cycle can also be changed at fixed pump power.

Starting from an SCD and appropriate adjusting of the PCs, dual color domain (DCD), consisting of two domains belonging to different wavelengths, coexist in the cavity and takes up nearly all the available space. The polarization-resolved analysis of the DCD indicates that they are partially polarized. In this analysis, the spectra are similar to those reported in references [27, 33, 40], and the temporal traces reveal that these adjacent polarization domains cannot be completely separated. However, the domains can be completely separated using an optical tunable filter (OTF). The output coupler (OC) immediately following the PD-ISO is used as the output port of the cavity to ensure that the generation of DCDs is not attributed to polarization switching. Outside the cavity, a PC and a polarizer were used to confirm that the output signal is linearly polarized. The same DCD phenomenon is observed in this situation, meaning that the wavelength switch is the key factor to form the DCDs. Figure 3.7a shows the typical optical spectrum of DCD operation. Figure 3.7b shows the DCD's temporal traces without a filter and filtered at short wavelength (SW) and long wavelength (LW), respectively. Two antiphase CDs generated by different wavelengths form the filtered signals. The duty cycle can be varied, as with SCD, by adjusting the orientation of the PCs. A proper adjustment of the controllers can lead to different wavelength spacing. The existence of an intensity dip (dark pulse) visible in the total intensity has been attributed to the combined effects of gain depletion [44] and chromatic dispersion-induced group velocity difference [31, 42]. PDW generation in fiber lasers is due to cross-polarization coupling through the optical Kerr effect, whereas DCD formation is attributed to cross-gain saturation through the amplifying medium.

Color domains can be created with more than two spectral components, unlike PDW, which is limited to two alternative domains associated with the two polarization eigenstates. Triple color domain (TCD) has been observed in reference [41], as illustrated in Figure 3.8. According to the polarization-resolved study, the three domains are also partially polarized. Figure 3.8a exhibits the temporal trace of TCD operation. The filtered signals at 1,608, 1,604, and 1,564 nm are shown in Figure 3.8b. The corresponding spectrum is presented in Figure 3.8c. The formation mechanism of TCD is also attributed to the cross-gain saturation. The duty cycle, amplitude, and wavelength spacing of TCD operation are all tunable in the same way as DCD is. In this particular regime, three-wavelength dark pulse and bright–dark pulse pair have been observed as well.

Additional physical insight is provided in reference [45]. Color domains in a ring cavity with a large anomalous dispersion are investigated. The cavity includes a polarization-sensitive isolator between two polarization controllers, and the length is

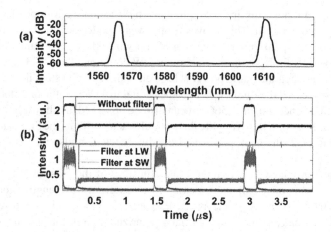

**FIGURE 3.7**  DCD characteristics. (a) Optical spectrum. (b) Temporal traces of initial and filtered signals. (from Meng et al. (2018) [41].)

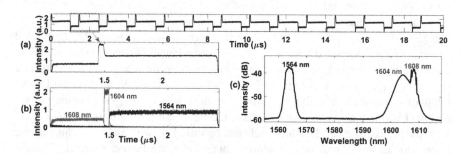

**FIGURE 3.8**  TCD characteristics. (a) Temporal trace. (b) Filtered temporal traces. (c) Optical spectrum. (from Meng et al. (2018) [41].)

long enough to prevent the emergence of solitons (about 100 m to 1 km). The effective Lyot effect results in a single, dual, or triple wavelength operation with tunable spacing, and single, dual, and triple color domains with well-separated wavelengths have been reported. In addition, harmonic mode-locking of SCD, DCD, and TCD has also been obtained, comparable to what happens in single- and dual-wavelength noise-like pulse emission [46]. Figure 3.9 shows the second-order harmonic operation of the laser. In the case of dual CD, the domains still exist with a considerable wavelength distance up to 60 nm. The wall separating the two domain manifests as a dark pulse and can be regarded as the presence of an attractive and repulsive force on both sides of the domain wall caused by the difference in group velocity between the two wavelengths [31, 42]. It should be noted that not all CDs have the same polarization state. The exact position of the output coupler plays a fundamental role. Indeed, the CDs have the same polarization states if the output is placed after the polarization-dependent isolator; however, if the light is extracted from another position, the CDs tend to have distinct polarizations. This behavior makes it difficult to distinguish

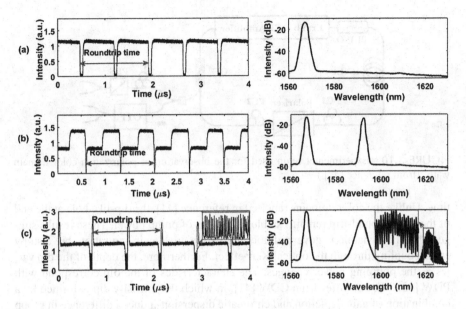

**FIGURE 3.9** Second-order harmonic operation of (a) single CD. (b) Dual CD and (c) Tricolor domain. Insets of (c) are zoomed temporal trace and spectrum, respectively. (from Meng et al. (2021) [45].)

between PD and CD. Therefore, the use of polarization-resolved or wavelength-resolved experiments is an excellent technique to determine the nature of the optical domain.

The existence of polarization-color domain walls, comprised of optical domains with well-defined polarization states and wavelength, was recently reported [47]. The optical configuration is given in Figure 3.10. It is a classical unidirectional ring cavity, including an intra-cavity polarizer and a piece of HNLF. The latter is used to increase the nonlinearity in terms of preventing the creation of conventional solitons while it enhances the cross-coupling which is beneficial to the formation of stable domain walls [40]. Multiwavelength operation is easily achieved thanks to the birefringent spectral filtering effect [48]. In contrast to previous works, the output signal is sent simultaneously to a polarization-resolved setup and a spectral tunable filter. The polarization controllers can be adjusted to achieve dual-wavelength operation, as shown in Figure 3.11a. As seen in Figure 3.11b, the output pulse train consists of a step-like pulse with an intensity dip (dark pulse). Wavelength-resolved measurements reveal that spectral components can be entirely separated and exhibit antiphase dynamics. This indicates that the output intensity is a combination of two-wavelength domain-wall group velocity locked, meaning they propagate as a unit without suffering from dispersion.

A further in-depth analysis indicates that as the pump power is increased, the optical spectrum remains almost constant while the duty cycle varies. In reference [41], the dark pulse width varied with pump power and was regarded as a gain recovery

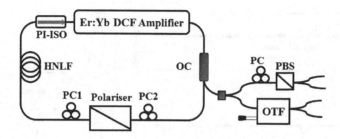

**FIGURE 3.10**  Experimental setup used for the observation of polarization-color domain walls. (from reference [47].)

time. Unlike the color domains reported in reference [41], the dip-like hole at the end of the cavity round-trip remains stable regardless of pump power. These results suggest that the dynamic is governed not only by the cross-gain saturation, but also by cross-coupling through the optical Kerr effect. Furthermore, the depth of the dip varies as the pumping power changes. The characteristics of the dip correspond with PDW [15, 30], but differ from CDW [41], in which the intensity dip is formed by a combination of gain depletion and chromatic dispersion-induced difference in group velocity. A polarization-resolved experiment was performed to confirm the existence of polarization domain walls. Figure 3.12 depicts the temporal evolution of the polarization components. It is demonstrated that the two polarization components are antiphase and travel as a unit, forming polarization domain walls. Remarkably, PDW are

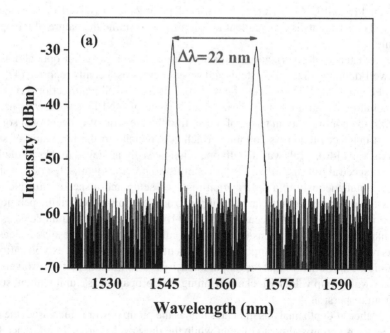

**FIGURE 3.11**  (a) The optical spectrum of the dual-wavelength operation.

(*Continued*)

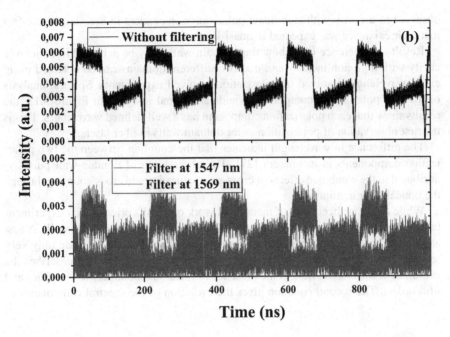

**FIGURE 3.11** (Continued) (b) Temporal evolution of the total intensity and spectrally filtered signals. (from reference [47].)

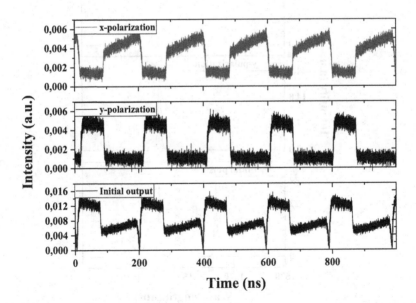

**FIGURE 3.12** Oscilloscope trace of the polarization-resolved experiment. (from reference [47].)

observed in a cavity with significant polarization-dependent losses, despite the fact that their existence was expected in quasi-isotropic cavities.

Results in reference [47] show that domain walls may be achieved in a dichroic cavity with high nonlinearity and that two different domains can be obtained using either wavelength-resolved or polarization-resolved experiments. Spectral analysis of the two polarization components yields additional insights. In Figure 3.13, the results show that each polarization component has a well-defined wavelength. This is the first observation of polarization-color domain walls in fiber lasers.

The difference in wavelength indicates that the coupling between the two polarization components is incoherent [27]. The observation of domain-wall pulses is attributed to the combined effects of cross-gain saturation and cross-coupling through the optical Kerr nonlinearity.

While there has been a lot of theoretical work on PDWs prior to their experimental observation, no theory has prompted the observation of color domain walls. A first attempt to model color domain wall formation in fiber lasers was made only very recently [49]. The model is based on two master equations. The first describes the evolution of the electric field in the presence of linear and nonlinear losses (third and fifth order). The second equation gives the evolution of the spectral components of

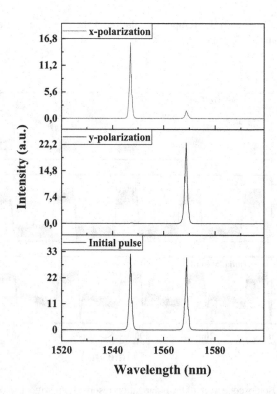

**FIGURE 3.13** Optical spectra of the two polarization components and the total intensity. (from reference [47].)

the electric field. It considers saturable gain and second-order dispersion. The former is assumed to be composed of two well-separated spectral areas saturated by total intensity and independently pumped. It is shown that color domains with identical amplitude can exist with the right set of parameters. A dark pulse forms a wall between the two domains as a result of the difference in GVD. It is also revealed that there are stair-case color pulses separated by a dip. This is the first theoretical model that allows the reproduction of color-domain formation in fiber lasers.

## 3.4 CONCLUSIONS

In conclusion, we reviewed recent experimental results on domain walls in fiber lasers. From a chronological standpoint, we first looked at the case of polarization domain walls. The results on color domain walls for which the domains are related with two separate wavelengths rather than two polarization components were then reported in detail. The fundamental differences are due to the physical mechanism by which they develop. The PDWs are caused by cross-coupling polarization caused by the optical Kerr effect, whereas the CDWs are caused by cross-gain saturation in the amplifying medium. We've also discussed polarization-color domain walls, in which each domain has its own polarization and wavelength. Domain walls appear as a common feature of fiber lasers.

## REFERENCES

[1] V.E. Zakharov and V. Mikhailov, "Polarization domains in nonlinear optics," *Sov. Phys. J.E.T.P. Lett. 45*, 349–352 (1987).

[2] M.V. Tratnik and S.E. Sipe, "Polarization solitons," *Phys. Rev. Lett. 58*, 1104–1107 (1987).

[3] D.N. Christodoulides and R.L. Joseph, "Vector solitons in birefringent nonlinear dispersive media," *Opt. Lett. 13*, 53–55 (1988).

[4] M.V. Tratnik and J.E. Sipe, "Bound solitary waves in a birefringent optical fiber," *Phys. Rev. A 38*, 2011–2017 (1988).

[5] D.N. Christodoulides, "Black and white vector solitons in weakly birefringent optical fibers," *Phys. Lett. A 132*, 451–452 (1988).

[6] D. David, D.D. Holm and M.V. Tratnik, "Hamiltonian chaos in nonlinear optical polarization dynamics," *Phys. Rep. 187*, 281–367 (1990).

[7] Y.S. Kivshar and S.K. Turitsyn, "Vector dark solitons," *Opt. Lett. 18*, 337–339 (1993).

[8] M. Haelterman, A.P. Sheppard and A.W. Snyder, "Bound-vector solitary waves in isotropic nonlinear dispersive media," *Opt. Lett. 18*, 1406–1408 (1993).

[9] A.V. Mikhailov and S. Wabnitz, "Polarization dynamics of counterpropagating beams in optical fibers," *Opt. Lett. 15*, 1055–1057 (1990).

[10] S. Wabnitz and B. Daino, "Polarization domains and instabilities in nonlinear optical fibers," *Phys. Lett. A 182*, 289–293 (1993).

[11] M. Haelterman and A.P. Sheppard, "Polarization domain walls in diffractive or dispersive Kerr media," *Opt. Lett. 19*, 96–98 (1994).

[12] M. Haelterman and M. Badolo, "Dual-frequency wall solitary waves for nonreturn to zero signal transmission in W-type single-mode fibers," *Opt. Lett. 20*, 2285–2287 (1995).

[13] S. Pitois, G. Millot and S. Wabnitz, "Polarization domain wall solitons with counterpropagating lasers beams," *Phys. Rev. Lett. 81*, 1409–1412 (1998).

[14] S. Pitois, G. Millot and S. Wabnitz, "Nonlinear polarization dynamics of counterpropagating waves in an isotropic optical fiber: theory and experiment," *JOSA B 18*, 432–443 (2001).

[15] S. Wabnitz, "Cross-polarization modulation domain wall solitons for WDM signals in birefringent optical fibers," *IEEE Photonics Technol. Lett. 21*, 875–877 (2009).

[16] M. Gilles, P.-Y. Bony, J. Garnier, A. Picozzi, M. Guasoni and J. Fatome, "Polarization domain walls in optical fibres as topological bits for data transmission," *Nat. Photon. 11*, 102–109 (2017).

[17] B. Garbin, J. Fatome, G.-L. Oppo, M. Erkintalo, S.G. Murdoch and S. Ceon, "Dissipative polarization domain walls in a passive coherently driven Kerr resonator," *Phys. Rev. Lett. 126*, 023904 (2021).

[18] S. Bielawski, B. Derozier and P. Glorieux, "Antiphase dynamics and polarization effects in the Nd-doped fiber laser," *Phys. Rev. A 46*, 2811–2822 (1992).

[19] B. Meziane, F. Sanchez, G.M. Stéphan and P.L. François, "Feedback-induced polarization switching in a Nd-doped fiber laser," *Opt. Lett. 19*, 1970–1972 (1993).

[20] A.J. Poustie, "Polarization cross saturation in an Er3+-doped fiber ring laser," *Opt. Lett. 20*, 1868–1970 (1995).

[21] Q.L. Williams and R. Roy, "Fast polarization dynamics of an erbium-doped fiber ring laser," *Opt. Lett. 21*, 1478–1480 (1996).

[22] P. Le Boudec, C. Jaouen, P.L. François, J.-F. Bayon, F. Sanchez, P. Besnard and G. Stéphan, "Antiphase dynamics and chaos in self-pulsing erbium-doped fiber lasers," *Opt. Lett. 18*, 1890–1892 (1993).

[23] F. Sanchez, M. Le Flohic, G. Stéphan, P. Le Boudec and P.L. François, "Quasi-periodic route to chaos in erbium-doped fiber laser," *IEEE J. Quant. Electron. 31*, 481–488 (1995).

[24] F. Sanchez and G. Stéphan, "General analysis of instabilities in erbium-doped fiber lasers," *Phys. Rev. E 53*, 2110–2122 (1996).

[25] H. Zhang, D.Y. Tang, L.M. Zhao and H.Y. Tam, "Induced solitons formed by cross-polarization coupling in a birefringent cavity fiber laser," *Opt. Lett. 33*, 2317–2319 (2008).

[26] H. Zhang, D.Y. Tang, L.M. Zhao and X. Wu, "Observation of polarization domain wall solitons in weakly birefringent cavity fiber lasers," *Phys. Rev. B 80*, 052302 (2009).

[27] H. Zhang, D.Y. Tang, L.M. Zhao and R.J. Knize, "Vector dark domain wall solitons in a fiber ring laser," *Opt. Express 18*, 4428–4433 (2010).

[28] D.Y. Tang, L.M. Zhao, B. Zhao and Q. Liu, "Mechanism of multisoliton formation and soliton energy quantization in passively mode-locked fiber lasers," *Phys. Rev. A 74*, 043815 (2005).

[29] C. Lecaplain, P. Grelu and S. Wabnitz, "Polarization domain wall complexes in fiber lasers," *JOSA B 30*, 211–218 (2013).

[30] C. Lecaplain, P. Grelu and S. Wabnitz, "Dynamics of the transition from polarization disorder to antiphase polarization domains in vector fiber lasers," *Phys. Rev. A 89*, 063812 (2014).

[31] D.Y. Tang, Y.F. Song, J. Guo, Y.J. Xiang and D.Y. Shen, "Polarization domain formation and domain dynamics in quasi-isotropic cavity fiber laser," *IEEE J. Sel. Top. Quantum Electron. 20*, 0901309 (2014).

[32] B. Malomed, "Domain walls between travelling waves," *Phys. Rev. E 50*, R3310–R3313 (1994).

[33] X. Li, S. Zhang, H. Han, M. Han, H. Zhang, L. Zhao, F. Wen and Z. Yang, "different polarization dynamic states in a vector Yb-doped fiber laser," *Opt. Express 23*, 10747–10755 (2015).

[34] J. Liu, X. Li, S. Zhang, H. Zhang, P. Yan, M. Han, Z. Pang and Z. Yang, "Polarization domain wall pulses in a microfiber-based topological insulator fiber laser," *Sci. Rep. 6*, 29128 (2016).

[35] W. Zhang, L. Zhan, T. Xian and L. Gao, "Harmonic mode-locking in bidirectional domain-wall soliton fiber lasers," *IEEE J. Light. Technol. 21*, 5417–5421 (2019).

[36] X. Hu, J. Guo, J. Ma, L. Li and D.Y. Tang, "Polarization domain splitting and incoherently coupled dark-bright vector soliton formation in single-mode fiber lasers," *J. Opt. Soc. Am. B 38*, 24–29, (2021).

[37] Y. Song, X. Shi, C. Wu, D.Y. Tang and H. Zhang, "Recent progress of study on optical solitons in fiber lasers," *Appl. Phys. Rev. 6*, 021313 (2019).

[38] H. Zhang, D.Y. Tang, L.M. Zhao and X. Wu, "Dual-wavelength domain wall solitons in a fiber ring laser," *Opt. Express 19*, 3525–3530 (2011).

[39] Z.B. Lin, A.P. Luo, S.K. Wang, H.Y. Wang, W.J. Cao, Z.C. Luo and W.C. Xu, "Generation of dual-wavelength domain-wall rectangular-shape pulses in HNLF-based fiber ring laser," *Opt. Laser Technol. 44*, 2260–2264 (2012).

[40] P. Wang, K. Zhao, X. Xiao and C. Yang, "Pulse dynamics of dual-wavelength dissipative soliton resonances and domain wall solitons in a Tm fiber laser with fiber-based Lyot filter," *Opt. Express 24*, 30708–30719 (2017).

[41] Y. Meng, G. Semaan, M. Kemel, M. Salhi, A. Komarov and F. Sanchez, "Color domains in fiber lasers," *Opt. Lett. 43*, 5054–5057 (2018).

[42] S. Chouli and P. Grelu, "Soliton rains in a fiber laser: an experimental study," *Phys. Rev. A 81*, 063829 (2010).

[43] A. Niang, F. Amrani, M. Salhi, P. Grelu and F. Sanchez, "Rains of solitons in a figure-of-eight passively mode-locked fiber laser," *Appl. Phys. B 116*, 771–775 (2014).

[44] A. Haboucha, H. Leblond, M. Salhi, A. Komarov and F. Sanchez, "Analysis of soliton pattern formation in passively mode-locked fiber lasers," *Phys. Rev. A 78*, 043806 (2008).

[45] Y. Meng, D. Zhang, G. Semaan, M. Kemel, A. Nady, M. Salhi, A. Komarov and F. Sanchez, "Optical domains in fiber lasers," *J. Opt. 23*, 035502 (2021).

[46] E. Bravo-Huerta, M. Duran-Sanchez, R.I. Alvarez-Tamayo, H. Santiago-Hernandez, M. Bello-Jimenez, B. Posada-Ramirez, B. Ibarra-Escamilla, O. Pottiez and E.A. Kuzin, "Single and dual-wavelength noise-like pulses with different shapes in a double-clad Er/Yb fiber laser," *Opt. Express 27*, 12349–12359 (2019).

[47] A. Nady, G. Semaan, M. Kemel, M. Salhi and F. Sanchez, "Polarization-color domain walls in fiber ring lasers," *J. Light. Technol 38*, 6905–6910 (2020).

[48] Z.C. Luo, A.P. Luo, W.C. Xu, H.S. Yin, J.R. Liu, Q. Ye and Z.J. Fang, "Tunable multiwavelength passively mode-locked fiber ring soliton laser using intracavity birefringence-induced comb filter," *IEEE Photon. J. 2*, 571 (2010).

[49] A. Komarov, A. Dmitriev, K. Komarov, L. Zhao and F. Sanchez, "Formation of color solitons and domains in fiber lasers," *Opt. Lett. 47*, 1029–1032 (2022).

# 4 Dual-Output Vector Soliton Fiber Lasers

## Michelle Y. Sander and Shutao Xu

## CONTENTS

4.1 Introduction ............................................................................................... 123
4.2 Dual Frequency Comb Generation ......................................................... 124
4.3 Dual Output Pulse Trains in Mode-Locked Fiber Lasers ..................... 124
    4.3.1 Bi-Directional Propagation and Cavity-Space Multiplexing ........... 125
    4.3.2 Spectral Properties and Dual-Wavelength Operation ...................... 127
4.4 Polarization-Multiplexed Dual Output Pulse Trains ............................. 128
    4.4.1 Fundamental Mechanism of Polarization Rotation Vector Solitons ...129
    4.4.2 Period-Doubled Dual-Polarization Output Pulse Trains .................. 129
4.5 Dynamics of Dual Output Pulse Trains .................................................. 133
4.6 Conclusions ............................................................................................... 134
Acknowledgements ............................................................................................ 135
References ........................................................................................................... 135

## 4.1 INTRODUCTION

In any mode-locked fiber laser exhibiting inherent fiber birefringence, vector solitons can be formed and propagate within the laser cavity, especially if no polarization constraining device is incorporated within the cavity. Depending on the overall fiber birefringence and polarization asymmetry in any laser cavity, different types of optical solitons (Song et al., 2019) can be generated, including scalar solitons (Sotor, Sobon, and Abramski, 2012) and vector solitons (Christodoulides, 1988; Christodoulides and Joseph 1988; Cundiff et al., 1999). For vector solitons two orthogonally polarized coupled modes exist as a pair and they can feature diverse time-varying polarization dynamics. Group-velocity-locked solitons (GVLS) (Zhao et al., 2008b; Zhao et al., 2010; Jin et al., 2016; Zhu et al., 2018) are typically characterized by a self-frequency induced wavelength shift between their underlying two orthogonal polarization components, so that the group velocity difference induced by fiber birefringence is balanced by fiber dispersion and enables soliton trapping (Zhao et al., 2008b). Thus, the two polarization eigenmodes can be locked to emit stable, elliptically polarized solitons that propagate as a single entity with the same group velocity. For polarization-locked vector solitons (PLVS) (Mou et al., 2011), the two orthogonal polarization components are phase-locked based on the interplay between nonlinear polarization rotation and the inherent birefringence. Polarization rotation vector solitons (PRVS) (Cundiff, Collings, and Knox, 1997; Akhmediev et al., 1998; Cundiff et al., 1999; Collings et al., 2000; Soto-Crespo et al., 2000;

DOI: 10.1201/9781003206767-4

**123**

Akosman et al., 2018) evolve with an overall rotating polarization ellipse, typically at multiples of the cavity round-trip time. Utilizing these unique vector soliton properties can be very attractive to obtain multi-output pulse trains of interest for optical frequency comb applications (Cundiff, 2007; Diddams, 2010; Fortier and Baumann, 2019). Optical frequency combs have provided a powerful technique with ultra-high precision and sensitivity that has formed the cornerstone for spectroscopy, frequency metrology, optical communications, distance measurements, arbitrary waveform generation, or astronomical spectrograph calibration.

While the fundamental concepts of different types of polarization vector solitons were first discussed several decades ago, this chapter focuses on how the attractive properties of vector solitons can be shaped to obtain dual output pulses and especially dual-polarization output pulse trains from a single laser cavity. A unique case of polarization rotation vector solitons operating in a period-doubling state (corresponding to the operation at half the fundamental repetition rate) are discussed.

## 4.2   DUAL FREQUENCY COMB GENERATION

The generation of dual optical frequency combs (Coddington, Swann, and Newbury, 2010; Newbury, Coddington, and Swann, 2010; Ideguchi et al., 2014; Coddington, Newbury, and Swann, 2016; Millot et al., 2016) has been of significant interest for high-precision metrology and spectroscopy applications. Due to the two pulse trains having a slightly different pulse repetition rate spacing, temporal asynchronous optical sampling can be pursued when the pulses overlap in a temporal interferometer so that in a heterodyne measurement beat notes are generated on a photodetector. The optical characteristics are thus down-converted into the ratio-frequency (RF) domain at high sampling speeds without requiring any mechanically moving optical parts. However, conventionally, such systems required two distinct frequency comb sources that were electronically stabilized to each other with good phase-locking for relative high coherence properties, resulting in a fairly complex setup. Thus, it is attractive to generate these two output pulses from a single laser oscillator, which overall reduces the optical components, complexity and feedback mechanism required. In addition, such dual optical frequency combs from a single laser cavity can operate with common-mode noise cancellation and a higher degree of mutual coherence between the pulses, offering overall a less bulky and more user-friendly operation. This opens a promising research field for dual-output compact single sources relying on various mechanisms such as cavity-space, wavelength and polarization multiplexing to generate two ultrafast output pulse trains. In this chapter, we review first the mechanisms to generate dual-pulsed outputs from fiber lasers before focusing specifically on the generation of polarization rotation vector solitons for polarization-multiplexed pulse trains.

## 4.3   DUAL OUTPUT PULSE TRAINS IN MODE-LOCKED FIBER LASERS

Generating multiple output pulse trains has been an attractive quest pursued to be found in any mode-locked laser system, from solid state lasers, for example in Nd:YAG (Thévenin, Vallet, and Brunel, 2012) or Ti:Sapphire (Ideguchi et al., 2016), external disc semiconductor lasers (Link et al., 2015; Link, Klenner, and Keller, 2016; Link et al., 2017), double external cavity with vertical-cavity surface-emitting lasers (Marconi et al., 2015), microresonators (Yang et al., 2017; Dutt et al., 2018; Xu et al., 2021), Er:ZBLAN glass lasers (Hébert et al., 2017; Hébert et al., 2018), and Cr:ZnS lasers (Vasilyev et al., 2021) to fiber laser systems. Such lasers can be very attractive not only for dual comb spectroscopy, but also for optical sensing or terahertz spectroscopy.

Various mechanisms have been utilized to generate two output pulse trains in fiber laser systems, relying on various degrees of freedom to initiate different pulse propagation regimes (Liao et al., 2020): Bi-directional propagation or the incorporation of non-common optical paths within a laser cavity (circulation-direction and cavity-space multiplexing), the generation of pulses with slightly different center wavelengths (wavelength multiplexing) or the utilization of polarization multiplexing exploiting fiber birefringence. Many of these first laser demonstrations occurred in erbium-doped (Er) fiber lasers, but they have now expanded to laser systems based on all the other fiber gain materials, including ytterbium (Yb), thulium (Tm), or bismuth (Bi).

### 4.3.1 BI-DIRECTIONAL PROPAGATION AND CAVITY-SPACE MULTIPLEXING

Some laser cavities rely on creating two slightly asymmetric optical paths for the fields propagating through the gain fiber in forward and backward direction, thus effectively generating a dual pulse output train. These cavities generally do not include any direction-constraining device (e.g., isolator or circulator), so that e.g., in a ring laser both clockwise and counterclockwise pulse generation can be supported. A bi-directional Er-doped passively mode-locked ring fiber laser with a repetition rate offset was first experimentally demonstrated in 2008 (Kieu and Mansuripur, 2008) and various implementations for bi-directional fiber laser cavities followed (Liao et al., 2020; Dai et al., 2021). Based on the cavity design and the degree of asymmetry between both paths, the output pulse characteristics can either remain fairly similar or offer distinct differences in repetition rate, pulse duration, optical spectrum, and center wavelength. Depending on the gain spectrum tuning, pulses with non-overlapping optical spectra (Zhao et al., 2014), pulse trains with slight offsets in their center wavelength (Zeng, Liu, and Yun, 2013) or dual output pulses with very similar wavelength characteristics in Er (Mehravar et al., 2016; Krylov et al., 2016) and Tm soliton fiber lasers (Olson et al., 2018), see Figure 4.1a, could be generated. Stretched pulse behavior (Yao, 2014) and bi-directional propagation of two-soliton bound states (Li et al., 2016), as well as conventional and dissipative solitons depending on the propagation direction, have been further demonstrated (Cui and Liu, 2013). In addition, in cavities where filtering based on a Sagnac loop was utilized, the pulses tended to travel in opposite directions as well (Yun et al., 2012; Li et al., 2018; Liao et al., 2018). Relying on hybrid mode-locking based on nonlinear polarization rotation and a carbon nanotube saturable absorber, conventional solitons, Gaussian, and stretched parabolic pulses were generated using a distributed polarizer in a bi-directional fiber cavity (Krylov et al., 2016). In a similar mode-locking regime, uni-directional as well as bi-directional propagation

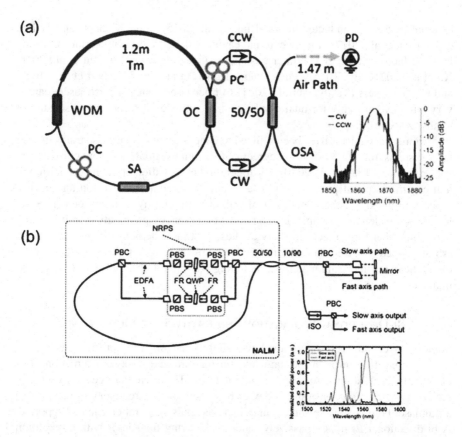

**FIGURE 4.1** (a) Bi-directional Tm-doped ring laser with clockwise and counter-clockwise propagating pulses with overlapping output spectra (Olson et al., 2018). (b) All PM Er-doped fiber laser with two separate gain fiber sections and nonlinear reciprocal phase shifters (NRPS) in a NALM loop with orthogonally polarized output pulses with different center wavelengths (adapted with permission from Figures 1 and 2 in Nakajima, Hata, and Minoshima, 2019 © The Optical Society).

based on the careful adjustment of the polarization state and fiber birefringence was presented (Chernysheva et al., 2016). Incorporating a nonreciprocal polarization iso-lator (combination of a polarizer with a 45° angle between two 45° Faraday rota-tors), an orthogonally-polarized bi-directional pulsed output (Uyama et al., 2022) was obtained. Other implementations relied on cavity-space multiplexing methods, for example, by using fiber loops that shared only a common gain fiber piece and mostly separate paths for each polarization (Kolano et al., 2018), or utilizing two dif-ferent saturable absorbers accessible via a four-port circulator (Ouyang et al., 2011; Mamidala et al., 2014) or two circulators (Saito et al., 2019) to differentiate between clockwise and counterclockwise pulses. Relying on a polarization rotation loop mir-ror with different paths for orthogonally polarized pulses from a linear gain section, polarization mode dispersion could be balanced, enabling the generation of synchro-nous pulses (Hu et al., 2020). Using a graded-index multimode fiber combined with bi-directional propagation in an Er-doped ring laser, different evolution trajectories for the asynchronous clockwise and counterclockwise pulses were observed at higher pump power values (Lin et al., 2022). While most of the demonstrated systems operated at more modest pulse energies, the pulse energy of dissipative solitons generated in a double-cladding Yb-doped gain fiber cavity in a bi-directional configuration was scaled to > 1 nJ in an all-normal dispersion cavity (Li et al., 2020).

An all-PM Er-doped fiber laser cavity based on a cavity-space multiplexing approach relied on two separate gain fiber sections and non-reciprocal phase shifters in a nonlinear amplifying loop mirror (NALM) (Nakajima, Hata, and Minoshima, 2019), thus creating two partially shared paths, cf. Figure 4.1b. Depending on the fine-tuning of the internal cavity losses, either a dual-wavelength operation (center wavelengths at 1,564.4 nm and 1,535.9 nm) with a difference in the repetition rate of 8.9 kHz (repetition rate of 21.2 MHz) or a single-wavelength polarization-multiplexed output (center wavelength of 1,562.7 nm) with a repetition rate offset of 9.5 kHz (repetition rate of 22.7 MHz) was achieved. The relative difference frequency beat note between both polarization components featured a linewidth around 1 kHz for a free-running operation, indicating the high relative coherence of the two generated output pulse trains (Nakajima, Hata, and Minoshima, 2019) and good robustness against environmental perturbations. Utilizing free-running bi-directional fiber laser sources, dual comb spectroscopy (Olson et al., 2018; Saito et al., 2019; Mizuno et al., 2021) as well as asynchronous optical sampling terahertz spectroscopy (Baker et al., 2018) has been demonstrated.

### 4.3.2 SPECTRAL PROPERTIES AND DUAL-WAVELENGTH OPERATION

In wavelength-multiplexed laser systems, two output pulse trains can simultaneously oscillate asynchronously in the cavity based on differences in the group velocity dispersion, while their optical spectra are characterized by distinct central wavelengths (see optical spectra in Figure 4.1b). These fiber laser cavities usually rely on an intra-cavity spectral filter that opens dual net gain windows for preferred simultaneous lasing operation. Combined with the wavelength-dependent fiber dispersion, this can then lead to a slight repetition rate offset. While fiber Bragg grating laser cavities have been commonly used for multi-wavelength continuous wave operation, they tend to support a narrower bandwidth of a few nanometers per spectral window and are thus not as optimized for ultrafast lasers (Park, Dawson, and Vahala, 1992; Li and Chan, 1999). Most commonly, a Lyot filter (a birefringence-induced filter) implementation can offer a sufficiently broad spectral bandwidth to enable the generation of ultrafast pulses. A Lyot filtering effect can be achieved through multiple mechanisms, frequently by incorporating a piece of polarization maintaining (PM) fiber (Zhao et al., 2018), combined with nonlinear polarization rotation or utilizing the induced birefringence from bending single-mode fiber (Zhu et al., 2019). Overall, a dual-wavelength ultrafast pulsed operation was demonstrated in all the various fiber laser systems, in Er-doped (Gong et al., 2006; Zhang et al. 2009a; Luo et al., 2010; Dong et al., 2011; Yun and Han, 2012; Liu et al., 2016; Wu et al., 2016; Zhao et al., 2016; Hu et al., 2017; Lin et al., 2017; Wei et al., 2018; Luo et al., 2019; Nakajima, Hata, and Minoshima, 2019; Zhu et al., 2019; Jin et al., 2020; Zhu et al., 2020),

Yb-doped (Zhang, Xu, and Zhang, 2012; Xu and Zhang, 2013; Huang et al., 2014; Fellinger et al., 2019), Bi-doped (Dong et al., 2011) and Tm-doped gain fiber laser cavities (Jiang et al., 2016; Wang et al., 2016; Liu et al., 2019; Xu et al., 2022b). While most systems operated in a soliton pulse regime in a cavity with net anomalous dispersion, dual-wavelength vector dissipative solitons in a net normal dispersion regime were generated in a similar manner. While most dual-wavelength operations led to an asynchronous dual output pulse generation, the pulses could also be synchronized when the group velocity difference induced by chromatic dispersion and polarization mode dispersion was well balanced (Zhang et al., 2009a). In an Er-doped fiber ring laser cavity with a piece of dispersion compensation fiber, a semiconductor saturable absorber mirror and an intracavity polarization controller, not only single- and dual-, but also triple-wavelength dissipative solitons were demonstrated (Zhang et al., 2009a). However, depending on the implementation and cavity design (e.g., by including a short section of PM fiber), two asynchronous pulse trains were generated with similar optical spectra that featured close-to-orthogonal states of polarization but not a completely linearly polarized pulse output (Zhao et al. 2018; Zhao et al. 2021). With an intracavity polarization controller the offset in the repetition rate could be tuned between 228 and 773 Hz (for a fundamental repetition rate around 44.1 MHz). Using this approach, asynchronous pulse trains with comparable center wavelengths and optical spectra were generated in a net anomalous soliton regime as well as in a net normal dispersion regime (the insertion of dispersion compensating fiber) for dissipative solitons (Zhao et al., 2018). The importance of a spectral band-pass filter to balance the gain competition and cross-phase modulation (Deng et al., 2019) in cross-polarized output pulses with overlapping spectra was demonstrated experimentally and confirmed by a theoretical model based on the complex Ginzburg-Landau quintic equation. Recently, nonlinear multimode interference in a single-mode fiber-graded-index multimode-single-mode fiber structure combined with a PM fiber laser enabled dual-wavelength mode-locking in an Er fiber laser (Zhao et al., 2019). In a similar approach with both Er-doped and Tm-doped gain fibers incorporated into the same laser cavity, simultaneous femtosecond pulse operation in those two separated bands (Zhao et al., 2020b) was shown. The free-running dual-wavelength mode-locked lasers have been successfully incorporated into various applications, particularly in dual comb spectroscopy (Zhao et al., 2016; Chen et al., 2019), for water detection from a single Tm-doped oscillator (Liao et al., 2018), gas analysis in a multipass-assisted gas sensor to characterize $C_2H_2/NH_3$ mixtures (Xu et al., 2020) and for combustion diagnostics (Xu et al., 2022a) or dual comb supercontinuum generation (Gu et al., 2021).

## 4.4 POLARIZATION-MULTIPLEXED DUAL OUTPUT PULSE TRAINS

By incorporating a birefringent medium into the laser cavity, two pulse trains with orthogonal polarization states can be generated within the same laser cavity based on the difference in the refractive index along each axis (between ordinary/extraordinary or fast/slow axis, respectively). One common option to tailor the birefringence is by including a short section of PM fiber into the laser cavity (X. Zhao et al., 2018; Deng et al., 2019; Zhao et al., 2019; Zhao et al., 2020a; Zhao et al., 2021;

Tao et al., 2022) and careful balancing could then lead to dual polarization output pulse trains with overlapping optical spectra (Zhao et al., 2018; Zhao et al., 2021; Tao et al., 2022), benefiting from a high intrinsic spectral coherence. Polarization-multiplexed dual output pulse trains tend to have a relatively small repetition rate offset that can be of the order of a few hundreds of Hertz. However, in general, any of the previously described mechanisms also allows the generation of polarization-multiplexed outputs and several of the cavity-space-multiplexed systems (Kolano et al., 2018; Nakajima, Hata, and Minoshima, 2019; Hu et al., 2020) or bi-directional configurations (Krylov et al., 2016; Uyama et al., 2022) led to dual polarized output pulses. Polarization-multiplexed dual-wavelength systems were also demonstrated for dissipative soliton cavities (Zhang et al., 2009a) or utilizing a nonlinear multi-mode interference effect for a saturable absorber (Zhao et al., 2019). For multiple vector solitons within the same cavity, the polarization dynamics become more complex: due to polarization-dependent gain competition, these vector solitons can either have orthogonal polarization orientations and propagate with the same group velocity or they can undergo polarization rotation in harmonically mode-locked vector solitons lasers (Akosman et al., 2018). Overall, orthogonally polarized output pulse trains can open up interesting application spaces where the polarization characteristic can be utilized to provide an additional degree of freedom or polarization sensitivity.

### 4.4.1 FUNDAMENTAL MECHANISM OF POLARIZATION ROTATION VECTOR SOLITONS

Polarization rotation vector solitons were first demonstrated in a linear soliton laser fiber cavity that was mode-locked with a semiconductor saturable absorber (Cundiff, Collings, and Knox, 1997; Cundiff et al., 1999). As the laser cavity in this instance did not feature any explicit polarizing element or polarization maintaining fiber, the polarization could evolve freely, influenced by the fiber birefringence (that can be strain- or bending-induced or associated with the slight asymmetry of the fiber core), nonlinear polarization rotation and cross-phase modulation. With an external linear polarizer in the output pulse train, the projection of the evolving polarization ellipse can be observed as an amplitude modulation (see Figure 4.2a). The polarization rotation dynamics of such vector solitons are overall governed by the cavity boundary conditions, which can lead to synchronization of the polarization rotation periodicity with the cavity length so that the rotation period tends to be an integer multiple of the cavity round-trip time. The resulting amplitude modulation can be measured in the frequency domain as a polarization evolution frequency (PEF), leading to sidebands in the main RF fundamental repetition rate and its harmonics, cf. Figure 4.2a.

### 4.4.2 PERIOD-DOUBLED DUAL-POLARIZATION OUTPUT PULSE TRAINS

While the PEF can be manipulated based on changes in the fiber birefringence, in particular through a polarization controller angle adjustment (Akosman and Sander, 2017), there exists one unique state where the PEF is locked at exactly half the fundamental repetition rate, leading to a period-doubled vector soliton, cf. Figure 4.2b. Such a period-doubled vector state was first demonstrated among different polarization rotation dynamics resolved in an Er-doped soliton fiber laser mode-locked

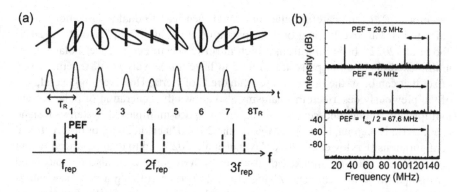

**FIGURE 4.2** (a) Illustration of the polarization rotation vector soliton characteristics after an external polarizer in the output pulse train: Output pulse-to-pulse evolution featuring a polarization ellipse projected into one polarization axis (top), the corresponding output pulse amplitude modulation with a periodicity over eight round trips (middle) and the RF spectrum highlighting the polarization evolution frequency (PEF) (bottom). (b) PEF tuning based on rotating the angle of an intracavity polarization controller in a linear Tm-doped gain fiber laser cavity with a fundamental repetition rate $f_{rep}$ of 135.2 MHz led to a unique period-doubled state with the PEF at half the initial repetition rate. (Adapted from reprinted with permission from Akosman and Sander, 2017 © The Optical Society.)

by a semiconductor saturable absorber mirror (Zhao, Tang, Zhang, and Wu, 2008a). Figure 4.3a shows the optical spectra of the period-doubled vector solitons before and after an external polarizer. The polarization-resolved spectra had comparable center wavelengths and featured characteristic peak-dip sideband pairs, which marks a signature of coherent energy exchange between the two polarization components induced by four-wave mixing (FWM). As the polarization ellipse of the vector soliton alternated with a period of twice the cavity round-trip time, the pulse train after the external polarizer featured an intensity difference between odd and even roundtrips, while the pulse train before the polarizer had a fairly constant intensity, as shown in Figure 4.3b.

Period-doubling (Akhmediev, Soto-Crespo, and Town, 2001), as one unique state for PRVS, has been demonstrated in different cavity net dispersion regimes and in diverse mode-locking scenarios, including soliton operation (Zhao, Tang, Zhang, and Wu, 2008a; Wong et al., 2011; Han et al., 2015; Wu et al., 2016; Akosman and Sander, 2017), stretched pulses (Yan et al., 2018) and dissipative solitons (Zhang et al., 2009b; Liu et al., 2017; Wang et al., 2020a). Although most of the research has been conducted in Er-doped fiber lasers, period-doubling dynamics have also been reported at other wavelength bands in Yb-doped lasers (Wong et al., 2011; Wang et al., 2020a) and Tm-doped lasers (Wu et al., 2016; Akosman and Sander, 2017). For most cases, the polarization-rotation vector solitons are elliptically polarized. As a consequence, the weaker pulse in the period-doubling states generated from those vector solitons could not be fully suppressed (Zhao, Tang, Zhang, and Wu, 2008a; Wong et al., 2011; Han et al., 2015; Wu et al., 2016; Liu et al., 2017; Yan et al., 2018). However, if the vector solitons were circularly polarized (Akosman and Sander, 2017; Wang et al., 2020a), both orthogonally polarized eigenstates had the same amplitude, leading to a unique state of period-doubling dynamics with ultra-high intensity modulation depth

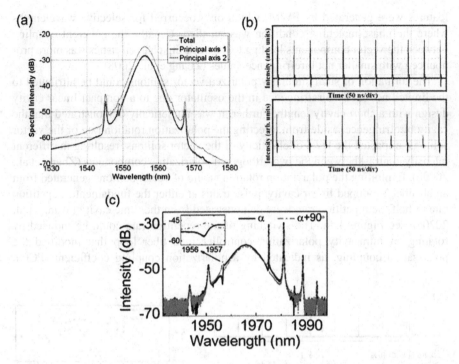

**FIGURE 4.3** (a) Output optical spectra before polarizer and after polarizer along two orthogonal principal polarization axes of a period-doubled vector soliton state. (b) Oscilloscope traces before (top) and after the polarizer (bottom), illustrating the period-doubling behavior after the external polarizer. (c) Optical spectra of a polarization-multiplexed period-doubled linear Tm-doped fiber laser for two orthogonal eigenstates at angle $\alpha$ and $\alpha+90°$, showing characteristic peak-dip polarization sidebands representative for coherent energy exchange. (reprinted with permission from Zhao, Tang, Zhang, and Wu, 2008a © The Optical Society.)

between odd and even roundtrips. While most cavity implementations relied on ring cavity lasers, a fully period-doubled system with linear output polarization in each state was implemented in a linear laser cavity based on Tm-doped gain fiber (Akosman and Sander, 2017). In such a soliton mode-locked laser with a semiconductor saturable Bragg reflector without any polarization selective fiber or device, varying an in-line polarization controller (PC) altered the net cavity birefringence and allowed the generation of PRVS with different PEFs (compare Figure 4.2b). For specific settings of the PC, circularly polarized vector solitons with period-doubling dynamics were obtained. Figure 4.3c shows the decomposition of the period-doubled PRVS into two polarization eigenstates, characterized by the strongest coherent energy exchange between the two polarization axes. In this state, a FWM-induced peak-dip pair around the wavelength of 1,956.5 nm was maximized. While this peak-dip signature might be visible for specific orthogonal polarization projections, which usually comprise the polarization eigenstates after an external polarizer, this feature might not be visible for a complementary polarization projection pair description (e.g., that are offset by 45°). Numerical simulations for polarization-locked vector solitons (Zhang et al., 2008; Wu et al., 2016; Fan et al., 2018) confirmed that these peak-dip features were generated by FWM, which only occurred for selective wavelengths where the phase-matching condition was satisfied. Usually, the coherent coupling between the vector components leading to the peak-dip characteristics was more pronounced with smaller net birefringence values (Zhang et al., 2008).

The formation of such circularly polarized vector solitons could be attributed to the limited net cavity birefringence in the oscillator due to a compact linear cavity design with a short cavity length. Further, it was numerically demonstrated that the cavity birefringence, aside from affecting the polarization rotation rate of the vector solitons, impacted the level of ellipticity of the vector solitons, resulting in different intensity contrast between the two orthogonal polarization components (Zhang et al., 2009b). Exploiting the polarization rotation nature of vector solitons generated from an all-fiber Yb-doped linear cavity, pulse trains at either the fundamental repetition rate or half the repetition rate were demonstrated from the same cavity (Wang et al. 2020a), see Figure 4.4a. The switching of the repetition rate could be induced by rotating an intracavity polarization controller in a fiber loop that modified the polarization coupling, as indicated by a polarization coupling coefficient (PCC).

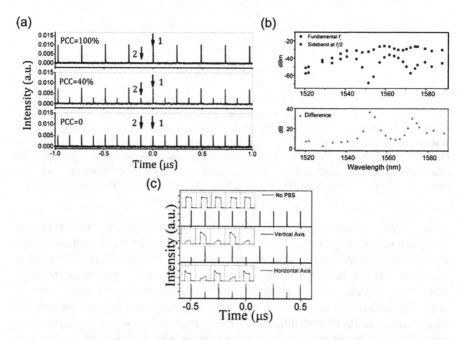

**FIGURE 4.4** (a) Switching between a period-doubled picosecond pulse train and pulses separated by the round-trip time based on a polarization coupling coefficient from a fiber loop led to the promotion of higher peak power and pulse energy for the frequency-halved state. (b) Wavelength dependency of the intensity modulation between adjacent roundtrips for a period-doubled state. (c) Shot-by-shot spectra of a period-doubled state, showing periodically alternating spectral characteristics after a polarization beam splitter (PBS) for the two polarization axes, as resolved with real-time DFT measurements. (reprinted with permission from Wang et al., 2020a © The Optical Society; reprinted with permission from Yan et al., 2018 © The Optical Society; reprinted with permission from Liu et al., 2017 © The Optical Society.)

For a maximized PCC, a repetition rate of 4.1 MHz was observed, which increased to the fundamental value of 8.2 MHz for PCC = 0. As the pulses experienced only a negligible change in their pulse duration, the period-doubling led to a beneficial promotion in pulse energy and peak power when the average power remained comparable. Thus, period-doubled polarization-rotation vector solitons can enable novel laser cavity designs and opportunities for repetition rate tuning as well as polarization multiplexing applications.

## 4.5 DYNAMICS OF DUAL OUTPUT PULSE TRAINS

For all the presented dual output pulse trains, the underlying spatio-temporal and spectral dynamics can be very interesting to study. Although period-doubling PRVS have been demonstrated in different laser cavities with diverse designs, they have been mostly characterized by standard instrumentation. By using a fast electro-optic modulator to pick pulses from either odd or even roundtrips from a period-doubled polarization-rotation-locked stretched-pulse fiber laser in both the temporal and spectral domains (Yan et al., 2018), the corresponding alternating features for the temporal profile and optical spectrum were revealed. As shown in Figure 4.4b, by spectrally selecting pulses with a tunable filter, the wavelength dependence of the period-doubling dynamics could be further studied. The intensity modulation depth between adjacent roundtrips varied with wavelengths and the strongest modulation occurred at the edge of the spectrum around 1,530 nm. Novel characterization methods focusing on the pulse-by-pulse dynamics have led to a deeper understanding of the period-doubling states of PRVS. With the development of real-time techniques like time-stretch dispersive Fourier transform (DFT) (Goda and Jalali, 2013; Wang et al., 2020b), the spectral characteristics of period-doubled polarization-rotation vector dissipative solitons in an Erbium-doped fiber laser with net-normal dispersion were studied roundtrip by roundtrip (Liu et al., 2017). Figure 4.4c presents both the oscilloscope traces and the shot-by-shot spectra captured with DFT before and after an external polarization beam splitter. Aside from emphasizing the alternating pulse intensity, the peak frequencies along the two orthogonal polarization axes dynamically alternated with a period of twice the cavity roundtrip time (Liu et al., 2017).

In general, the application of real-time measurements can offer interesting insights into the pulse evolution of asynchronous dual output pulse trains. Utilizing DFT, the collision dynamics between the two pulse polarization components in an Er-doped dual-output laser system were elucidated when the optical spectra of the temporal vector solitons with slightly different repetition rates either spectrally overlapped with each other (Zhao et al., 2021) or formed dual-color solitons (Wei et al., 2018). The polarization-dependent gain led to different buildup times for the two solitons (Zhao, Gao, et al., 2020a) and moderate or extreme collisions were observed when the pulses crossed each other temporally (Wei et al., 2018; Zhao et al., 2021). When phase-matching for both polarization components was obtained, FWM sidebands were generated, leading eventually to extreme collisions. Otherwise, the formation of weak pulses due to cross-polarization coupling was observed, together with a shift in the central wavelength in opposite directions after the collision, followed by reduced soliton interactions, intensity redistribution, and recovery. Pulse-by-pulse dual-color soliton spectral dynamics in a nonlinear-polarization rotation mode-locked Er fiber laser (Wei et al., 2018) revealed a collision-induced soliton self-reshaping process, characterized by dynamic fringes in the optical spectrum of the soliton and the rebuilding of Kelly sidebands with wavelength drifting after the collision. Numerical simulations confirmed that strong interactions based on cross-phase modulation during the collision process triggered a dispersive-wave shedding accompanied with soliton reshaping (Wei et al., 2018). For a dual-wavelength Yb-doped fiber laser, periodic explosions associated with spectral collapse and energy exchange between both solitons were observed before the pulses were restored to their initial intensities due to self-organization (Liu et al., 2020). Even the generation of dissipative rogue waves (Lecaplain et al., 2012) was confirmed for this laser configuration (Liu et al., 2020). In a transitionary pulsing state for a Tm fiber laser, instabilities and explosions, including the formation of period-doubled pulses (Zeng et al., 2022), were characterized with an adapted DFT system. For this asynchronous polarization rotation dual-output vector soliton laser, periodically occurring soliton explosions associated with spectral collapse were resolved and an enhanced pulse interaction period where the two pulses experienced a shift of their center wavelengths due to cross-phase modulation was identified. A damped multi-period modulation was observed as the pulse energy was dissipated before reaching equilibrium, indicative of the energy exchange between the two polarization components due to coherent coupling (Zeng et al., 2022). Thus, studying the real-time behavior of these dual-output lasers can offer novel insights into the rich underlying fundamental soliton dynamics as well as the governing nonlinearities, collisions, energy exchange, and instabilities of interest for novel laser designs and their application in precision spectroscopy and sensing.

## 4.6　CONCLUSIONS

We have reviewed the recent progress on different cavity configurations and mechanisms associated with the generation of two distinct mode-locked output pulses from a single fiber laser. Such sources with a slight offset in repetition rate are of particular interest for the emerging area of dual comb spectroscopy, where common mode suppression and thus enhanced noise performance combined with reduced oscillator complexity can offer some powerful advantages. Different fiber laser cavity configurations, including bi-directional propagation in ring lasers for circulation-direction multiplexing, the partial overlap of common paths within the cavity (cavity-space multiplexing), wavelength multiplexing as well as polarization multiplexing operation were presented as different routes to generate dual output pulse trains. Different mode-locking operations for solitons, stretched pulses, or dissipative solitons were discussed and cavity configurations incorporating various types of saturable absorbers in the form of semiconductors, graphene, carbon-nanotubes, nonlinear polarization evolution or nonlinear multi-mode interference were presented. A unique state of period-doubled polarization rotation vector solitons was discussed in detail since this regime offers frequency agility between the fundamental repetition rate as well as half its value. While most design and development efforts have been focused on shorter near-infrared wavelengths (primarily Er-doped fiber lasers or Yb-doped systems), recently a trend toward extending operating regimes to longer wavelengths has been pursued and some Tm-doped fiber laser dual-output systems were demonstrated. While silica-based fibers can readily support wavelengths up to slightly above 2 μm, mid-infrared fiber laser systems have been implemented and could offer interesting sources for dual comb spectroscopy in a spectral regime that features the most characteristic vibrational bands. As frequency diversity, multi-wavelength operation, polarization multiplexing and dual frequency comb operation become more interesting for various applications, novel implementations of dual-output fiber laser systems and studies of their spatio-temporal and spectral dynamics with real-time measurements can be regarded as a promising and enabling research field in the future.

## ACKNOWLEDGEMENTS

The authors would like to acknowledge funding from the National Science Foundation (ECCS-1710849) and the National Institute of Neurological Disorders and Stroke of the NIH (UF1NS107705).

## REFERENCES

Akhmediev, N. N., J. M. Soto-Crespo, S. T. Cundiff, B. C. Collings, and W. H. Knox. 1998. "Phase locking and periodic evolution of solitons in passively mode-locked fiber lasers with a semiconductor saturable absorber." *Optics Letters* 23 (11): 852–854. doi:10.1364/OL.23.000852.

Akhmediev, N., J. M. Soto-Crespo, and G. Town. 2001. "Pulsating solitons, chaotic solitons, period doubling, and pulse coexistence in mode-locked lasers: Complex Ginzburg-Landau equation approach." *Physical Review E* 63 (5): 056602. doi:10.1103/PhysRevE.63.056602.

Akosman, A. E., and M. Y. Sander. 2017. "Dual comb generation from a mode-locked fiber laser with orthogonally polarized interlaced pulses." *Optics Express* 25 (16): 18592–18602. doi:10.1364/OE.25.018592.

Akosman, A. E., J. Zeng, P. D. Samolis, and M. Y. Sander. 2018. "Polarization rotation dynamics in harmonically mode-locked vector soliton fiber lasers." *IEEE Journal on Selected Topics in Quantum Electronics* 24 (3): 1–7. doi:10.1109/JSTQE.2017.2768319.

Baker, R. D., N. T. Yardimci, Y.-H. Ou, K. Kieu, and M. Jarrahi. 2018. "Self-triggered asynchronous optical sampling terahertz spectroscopy using a bidirectional mode-locked fiber laser." *Scientific Reports* 8 (1): 14802. doi:10.1038/s41598-018-33152-0.

Chen, J., X. Zhao, Z. Yao, T. Li, Q. Li, S. Xie, J. Liu, and Z. Zheng. 2019. "Dual-comb spectroscopy of methane based on a free-running erbium-doped fiber laser." *Optics Express* 27 (8): 11406–11412. doi:10.1364/OE.27.011406.

Chernysheva, M., M. Al Araimi, H. Kbashi, R. Arif, S. V. Sergeyev, and A. Rozhin. 2016. "Isolator-free switchable uni- and bidirectional hybrid mode-locked erbium-doped fiber laser." *Optics Express* 24 (14): 15721–15729. doi:10.1364/OE.24.015721.

Christodoulides, D. N. 1988. "Black and white vector solitons in weakly birefringent optical fibers." *Physics Letters A* 132 (8): 451–452. doi:10.1016/0375-9601(88)90511-7.

Christodoulides, D. N., and R. I. Joseph. 1988. "Vector solitons in birefringent nonlinear dispersive media." *Optics Letters* 13 (1): 53–55. doi:10.1364/OL.13.000053.

Coddington, I., W. C. Swann, and N. R. Newbury. 2010. "Coherent dual-comb spectroscopy at high signal-to-noise ratio." *Physical Review A* 82 (4): 043817. doi:10.1103/PhysRevA.82.043817.

Coddington, I., N. Newbury, and W. Swann. 2016. "Dual-comb spectroscopy." *Optica* 3 (4): 414–426. doi:10.1364/OPTICA.3.000414.

Collings, B. C., S. T. Cundiff, N. N. Akhmediev, J. M. Soto-Crespo, K. Bergman, and W. H. Knox. 2000. "Polarization-locked temporal vector solitons in a fiber laser: Experiment." *JOSA B* 17 (3): 354–365. doi:10.1364/JOSAB.17.000354.

Cui, Y., and X. Liu. 2013. "Graphene and nanotube mode-locked fiber laser emitting dissipative and conventional solitons." *Optics Express* 21 (16): 18969–18974. doi:10.1364/OE.21.018969.

Cundiff, S. T., B. C. Collings, and W. H. Knox. 1997. "Polarization locking in an isotropic, modelocked soliton Er/Yb fiber laser." *Optics Express* 1 (1): 12–20. doi:10.1364/OE.1.000012.

Cundiff, S. T., B. C. Collings, N. N. Akhmediev, J. M. Soto-Crespo, K. Bergman, and W. H. Knox. 1999. "Observation of polarization-locked vector solitons in an optical fiber." *Physical Review Letters* 82 (20): 3988–3991. doi:10.1103/PhysRevLett.82.3988.

Cundiff, S. T. 2007. "Metrology: New generation of combs." *Nature* 450 (7173): 1175–1176. doi:10.1038/4501175b.

Dai, L., Z. Huang, Q. Huang, C. Zhao, A. Rozhin, S. Sergeyev, M. Al Araimi, and C. Mou. 2021. "Carbon nanotube mode-locked fiber lasers: Recent progress and perspectives." *Nanophotonics* 10 (2): 749–775. doi:10.1515/nanoph-2020-0446.

Deng, Z., Y. Liu, C. Ouyang, W. Zhang, C. Wang, and W. Li. 2019. "Mutually coherent dual-comb source generated from a free-running linear fiber laser." *Results in Physics* 14 (September): 102364. doi:10.1016/j.rinp.2019.102364.

Diddams, S. A. 2010. "The evolving optical frequency comb [Invited]." *Journal of the Optical Society of America B* 27 (11): B51–B62. doi:10.1364/JOSAB.27.000B51.

Dong, J.-L., X. Wen-Cheng, Z.-C. Luo, A.-P. Luo, H.-Y. Wang, W.-J. Cao, and L.-Y. Wang. 2011. "Tunable and switchable dual-wavelength passively mode-locked fiber ring laser with high-energy pulses at a sub-100kHz repetition rate." *Optics Communications* 284 (24): 5719–5722. doi:10.1016/j.optcom.2011.07.078.

Dutt, A., C. Joshi, X. Ji, J. Cardenas, Y. Okawachi, K. Luke, A. L. Gaeta, and M. Lipson. 2018. "On-chip dual-comb source for spectroscopy." *Science Advances* 4 (3): e1701858. doi:10.1126/sciadv.1701858.

Fan, X. L., Y. Wang, W. Zhou, X. J. Ren, Z. F. Hu, Q. Y. Liu, R. J. Zhang, and D. Y. Shen. 2018. "Generation of polarization-locked vector solitons in mode-locked thulium fiber laser." *IEEE Photonics Journal* 10 (1): 1–8. doi:10.1109/JPHOT.2017.2789302.

Fellinger, J., J. Fellinger, A. S. Mayer, G. Winkler, W. Grosinger, G.-W. Truong, S. Droste, et al. 2019. "Tunable dual-comb from an all-polarization-maintaining single-cavity dual-color yb: Fiber laser." *Optics Express* 27 (20): 28062–28074. doi:10.1364/OE.27.028062.

Fortier, T., and E. Baumann. 2019. "20 years of developments in optical frequency comb technology and applications." *Communications Physics* 2 (1): 1–16. doi:10.1038/s42005-019-0249-y.

Goda, K., and B. Jalali. 2013. "Dispersive fourier transformation for fast continuous single-shot measurements." *Nature Photonics* 7 (2): 102–112. doi:10.1038/nphoton.2012.359.

Gong, Y. D., X. L. Tian, M. Tang, P. Shum, M. Y. W. Chia, V. Paulose, J. Wu, and K. Xu. 2006. "Generation of dual wavelength ultrashort pulse outputs from a passive mode locked fiber ring laser." *Optics Communications* 265 (2): 628–631. doi:10.1016/j.optcom.2006.04.065.

Gu, X.-R., H.-D. Chen, Y. Li, X.-R. Cao, C.-S. Wang, and Y.-W. Liu. 2021. "Ultrashort pulse duration and broadband dual-comb laser system based on a free-running passively mode-locked Er-fiber oscillator." *Laser Physics Letters* 18 (12): 125101. doi:10.1088/1612-202X/ac308e.

Han, M., S. Zhang, X. Li, H. Zhang, H. Yang, and T. Yuan. 2015. "Polarization dynamic patterns of vector solitons in a graphene mode-locked fiber laser." *Optics Express* 23 (3): 2424–2435. doi:10.1364/OE.23.002424.

Hébert, N. B., J. Genest, J.-D. Deschênes, H. Bergeron, G. Y. Chen, C. Khurmi, and D. G. Lancaster. 2017. "Self-corrected chip-based dual-comb spectrometer." *Optics Express* 25 (7): 8168–8179. doi:10.1364/OE.25.008168.

Hébert, N. B., D. G. Lancaster, V. Michaud-Belleau, G. Y. Chen, and J. Genest. 2018. "Highly coherent free-running dual-comb chip platform." *Optics Letters* 43 (8): 1814–1817. doi:10.1364/OL.43.001814.

Hu, G., Y. Pan, X. Zhao, S. Yin, M. Zhang, and Z. Zheng. 2017. "Asynchronous and synchronous dual-wavelength pulse generation in a passively mode-locked fiber laser with a mode-locker." *Optics Letters* 42 (23): 4942–4945. doi:10.1364/OL.42.004942.

Hu, G., Y. Liu, X. Zhao, and Z. Zheng. 2020. "Polarization-multiplexed, synchronous ultrashort pulse generation from a linear-cavity fiber laser with a polarization-rotation loop mirror." *Optik* 224 (December): 165647. doi:10.1016/j.ijleo.2020.165647.

Huang, S., Y. Wang, P. Yan, J. Zhao, H. Li, and R. Lin. 2014. "Tunable and switchable multiwavelength dissipative soliton generation in a graphene oxide mode-locked Yb-doped fiber laser." *Optics Express* 22 (10): 11417. doi:10.1364/OE.22.011417.

Ideguchi, T., A. Poisson, G. Guelachvili, N. Picqué, and T. W. Hänsch. 2014. "Adaptive real-time dual-comb spectroscopy." *Nature Communications* 5 (February): 3375. doi:10.1038/ncomms4375.

Ideguchi, T., T. Nakamura, Y. Kobayashi, and K. Goda. 2016. "Kerr-lens mode-locked bidirectional dual-comb ring laser for broadband dual-comb spectroscopy." *Optica* 3 (7): 748–753. doi:10.1364/OPTICA.3.000748.

Jiang, K., W. Zhichao, F. Songnian, J. Song, H. Li, M. Tang, P. Shum, and D. Liu. 2016. "Switchable dual-wavelength mode-locking of thulium-doped fiber laser based on SWNTs." *IEEE Photonics Technology Letters* 28 (19): 2019–2022. doi:10.1109/LPT.2016.2580604.

Jin, X. X., Z. C. Wu, L. Li, Q. Zhang, D. Y. Tang, D. Y. Shen, S. N. Fu, D. M. Liu, and L. M. Zhao. 2016. "Manipulation of group-velocity-locked vector solitons from fiber lasers." *IEEE Photonics Journal* 8 (2): 1–6. doi:10.1109/JPHOT.2016.2545648.

Jin, X., M. Zhang, G. Hu, G. Hu, Q. Wu, Z. Zheng, Z. Zheng, and T. Hasan. 2020. "Broad bandwidth dual-wavelength fiber laser simultaneously delivering stretched pulse and dissipative soliton." *Optics Express* 28 (5): 6937–6944. doi:10.1364/OE.385142.

Kieu, K., and M. Mansuripur. 2008. "All-fiber bidirectional passively mode-locked ring laser." *Optics Letters* 33 (1): 64–66. doi:10.1364/OL.33.000064.

Kolano, M., O. Boidol, D. Molter, and G. Von Freymann. 2018. "Single-laser, polarization-controlled optical sampling system." *Optics Express* 26 (23): 30338–30346. doi:10.1364/OE.26.030338.

Krylov, A. A., D. S. Chernykh, N. R. Arutyunyan, V. V. Grebenyukov, A. S. Pozharov, and E. D. Obraztsova. 2016. "Generation regimes of bidirectional hybridly mode-locked ultrashort pulse erbium-doped all-fiber ring laser with a distributed polarizer." *Applied Optics* 55 (15): 4201. doi:10.1364/AO.55.004201.

Lecaplain, C., Ph. Grelu, J. M. Soto-Crespo, and N. Akhmediev. 2012. "Dissipative rogue waves generated by chaotic pulse bunching in a mode-locked laser." *Physical Review Letters* 108 (23): 233901. doi:10.1103/PhysRevLett.108.233901.

Li, S., and K. T. Chan. 1999. "A novel configuration for multiwavelength actively mode-locked fiber lasers using cascaded fiber Bragg gratings." *IEEE Photonics Technology Letters* 11 (2): 179–181. doi:10.1109/68.740696.

Li, L., Q. Ruan, R. Yang, L. Zhao, and Z. Luo. 2016. "Bidirectional operation of 100 fs bound solitons in an ultra-compact mode-locked fiber laser." *Optics Express* 24 (18): 21020. doi:10.1364/OE.24.021020.

Li, R., H. Shi, H. Tian, Y. Li, B. Liu, Y. Song, and M. Hu 2018. "All-polarization-maintaining dual-wavelength mode-locked fiber laser based on Sagnac loop filter." *Optics Express* 26 (22): 28302–28311. doi:10.1364/OE.26.028302.

Li, B., B. Li, J. Xing, D. Kwon, Y. Xie, N. Prakash, J. Kim, and S.-W. Huang. 2020. "Bidirectional mode-locked all-normal dispersion fiber laser." *Optica* 7 (8): 961–964. doi:10.1364/OPTICA.396304.

Liao, R., Y. Song, W. Liu, H. Shi, L. Chai, and M. Hu. 2018. "Dual-comb spectroscopy with a single free-running thulium-doped fiber laser." *Optics Express* 26 (8): 11046–11054. doi:10.1364/OE.26.011046.

Liao, R., H. Tian, W. Liu, R. Li, Y. Song, and M. Hu. 2020. "Dual-comb generation from a single laser source: Principles and spectroscopic applications towards mid-IR—A review." *Journal of Physics: Photonics* 2 (4): 042006. doi:10.1088/2515-7647/aba66e.

Lin, B., X. Zhao, M. He, Y. Pan, J. Chen, S. Cao, Y. Lin, Q. Wang, Z. Zheng, and Z. Fang. 2017. "Dual-comb absolute distance measurement based on a dual-wavelength passively mode-locked laser." *IEEE Photonics Journal* 9 (6): 1–8. doi:10.1109/JPHOT.2017.2772292.

Lin, J., Z. Dong, T. Dong, X. Ma, C. Dai, P. Yao, C. Gu, and L. Xu. 2022. "Bidirectional mode-locked fiber laser based on nonlinear multimode interference." *Optics & Laser Technology* 154 (October): 108269. doi:10.1016/j.optlastec.2022.108269.

Link, S. M., A. Klenner, M. Mangold, C. A. Zaugg, M. Golling, B. W. Tilma, and U. Keller. 2015. "Dual-comb modelocked laser." *Optics Express* 23 (5): 5521–5531. doi:10.1364/OE.23.005521.

Link, S. M., A. Klenner, and U. Keller. 2016. "Dual-comb modelocked lasers: Semiconductor saturable absorber mirror decouples noise stabilization." *Optics Express* 24 (3): 1889–1902. doi:10.1364/OE.24.001889.

Link, S. M., D. J. H. C. Maas, D. Waldburger, and U. Keller. 2017. "Dual-comb spectroscopy of water vapor with a free-running semiconductor disk laser." *Science* 356 (6343): 1164–1168. doi:10.1126/science.aam7424.

Liu, Y., X. Zhao, G. Hu, C. Li, B. Zhao, and Z. Zheng. 2016. "Unidirectional, dual-comb lasing under multiple pulse formation mechanisms in a passively mode-locked fiber ring laser." *Optics Express* 24 (19): 21392–21398. doi:10.1364/OE.24.021392.

Liu, M., A.-P. Luo, Z.-C. Luo, and W.-C. Xu. 2017. "Dynamic trapping of a polarization rotation vector soliton in a fiber laser." *Optics Letters* 42 (2): 330–333. doi:10.1364/OL.42.000330.

Liu, S., Z. Zhang, J. Shen, and K. Yu. 2019. "Wavelength-spacing adjustable dual-wavelength dissipative soliton resonance thulium-doped fiber laser." *IEEE Photonics Journal* 11 (2): 1–9. doi:10.1109/JPHOT.2019.2899618.

Liu, M., T.-J. Li, A.-P. Luo, W.-C. Xu, and Z.-C. Luo. 2020. "'Periodic' soliton explosions in a dual-wavelength mode-locked Yb-doped fiber laser." *Photonics Research* 8 (3): 246–251. doi:10.1364/PRJ.377966.

Luo, Z.-C., A.-P. Luo, W.-C. Xu, H.-S. Yin, J.-R. Liu, Q. Ye, and Z.-J. Fang. 2010. "Tunable multiwavelength passively mode-locked fiber ring laser using intracavity birefringence-induced comb filter." *IEEE Photonics Journal* 2 (4): 571–577. doi:10.1109/JPHOT.2010.2051023.

Luo, X., T. H. Tuan, T. S. Saini, H. P. T. Nguyen, T. Suzuki, and Y. Ohishi. 2019. "Tunable and switchable all-fiber dual-wavelength mode locked laser based on Lyot filtering effect." *Optics Express* 27 (10): 14635–14647. doi:10.1364/OE.27.014635.

Mamidala, V., R. I. Woodward, Y. Yang, H. H. Liu, and K. K. Chow. 2014. "Graphene-based passively mode-locked bidirectional fiber ring laser." *Optics Express* 22 (4): 4539–4546. doi:10.1364/OE.22.004539.

Marconi, M., J. Javaloyes, S. Barland, S. Balle, and M. Giudici. 2015. "Vectorial dissipative solitons in vertical-cavity surface-emitting lasers with delays." *Nature Photonics* 9 (7): 450–455. doi:10.1038/nphoton.2015.92.

Mehravar, S., R. A. Norwood, N. Peyghambarian, and K. Kieu. 2016. "Real-time dual-comb spectroscopy with a free-running bidirectionally mode-locked fiber laser." *Applied Physics Letters* 108 (23): 231104. doi:10.1063/1.4953400.

Millot, G., S. Pitois, M. Yan, T. Hovhannisyan, A. Bendahmane, T. W. Haensch, and N. Picque. 2016. "Frequency-agile dual-comb spectroscopy." *Nature Photonics* 10 (1): 27–U37. doi:10.1038/NPHOTON.2015.250.

Mizuno, T., T. Mizuno, T. Mizuno, Y. Nakajima, Y. Nakajima, Y. Nakajima, Y. Nakajima, et al. 2021. "Computationally image-corrected dual-comb microscopy with a free-running single-cavity dual-comb fiber laser." *Optics Express* 29 (4): 5018–5032. doi:10.1364/OE.415242.

Mou, C., S. Sergeyev, A. Rozhin, and S. Turistyn. 2011. "All-fiber polarization locked vector soliton laser using carbon nanotubes." *Optics Letters* 36 (19): 3831–3833. doi:10.1364/OL.36.003831.

Nakajima, Y., Y. Hata, and K. Minoshima. 2019. "All-polarization-maintaining, polarization-multiplexed, dual-comb fiber laser with a nonlinear amplifying loop mirror." *Optics Express* 27 (10): 14648. doi:10.1364/OE.27.014648.

Newbury, N. R., I. Coddington, and W. Swann. 2010. "Sensitivity of coherent dual-comb spectroscopy." *Optics Express* 18 (8): 7929–7945. doi:10.1364/OE.18.007929.

Olson, J., Y. H. Ou, A. Azarm, and K. Kieu. 2018. "Bi-directional mode-locked thulium fiber laser as a single-cavity dual-comb source." *IEEE Photonics Technology Letters* 30 (20): 1772–1775. doi:10.1109/LPT.2018.2868940.

Ouyang, C., P. Shum, K. Wu, J. H. Wong, H. Q. Lam, and S. Aditya. 2011. "Bidirectional passively mode-locked soliton fiber laser with a four-port circulator." *Optics Letters* 36 (11): 2089–2091. doi:10.1364/OL.36.002089.

Park, N., J. W. Dawson, and K. J. Vahala. 1992. "Multiple wavelength operation of an erbium-doped fiber laser." *IEEE Photonics Technology Letters* 4 (6): 540–541. doi:10.1109/68.141960.

Saito, S., M. Yamanaka, Y. Sakakibara, E. Omoda, H. Kataura, and N. Nishizawa. 2019. "All-polarization-maintaining er-doped dual comb fiber laser using single-wall carbon nanotubes." *Optics Express* 27 (13): 17868–17875. doi:10.1364/OE.27.017868.

Song, Y., X. Shi, C. Wu, D. Tang, and H. Zhang. 2019. "Recent progress of study on optical solitons in fiber lasers." *Applied Physics Reviews* 6 (2): 021313. doi:10.1063/1.5091811.

Soto-Crespo, J. M., N. N. Akhmediev, B. C. Collings, S. T. Cundiff, K. Bergman, and W. H. Knox. 2000. "Polarization-locked temporal vector solitons in a fiber laser: Theory." *JOSA B* 17 (3): 366–372. doi:10.1364/JOSAB.17.000366.

Sotor, J., G. Sobon, and K. M. Abramski. 2012. "Scalar soliton generation in all-polarization-maintaining, graphene mode-locked fiber laser." *Optics Letters* 37 (11): 2166–2168. doi:10.1364/OL.37.002166.

Tao, J., Q. Lin, L. Yan, L. Hou, B. Lu, and J. Bai. 2022. "Asynchronous vector solitons based dual-comb in a fiber laser mode-locked by GO-COOH SA." *Optics & Laser Technology* 154 (October): 108308. doi:10.1016/j.optlastec.2022.108308.

Thévenin, J., M. Vallet, and M. Brunel. 2012. "Dual-polarization mode-locked Nd: YAG laser." *Optics Letters* 37 (14): 2859. doi:10.1364/OL.37.002859.

Uyama, K., T. Shirahata, S. Y. Set, and S. Yamashita. 2022. "Orthogonally-Polarized bi-directional dual-comb fiber laser." In *Conference on Lasers and Electro-Optics (2022), Paper SF4H.4*, SF4H.4. Optica Publishing Group. doi:10.1364/CLEO_SI.2022.SF4H.4.

Vasilyev, S., M. Y. Sander, J. Gu, V. Smolski, I. Moskalev, M. Mirov, Y. Barnakov, et al. 2021. "Vector solitons in a kerr-lens mode-locked laser oscillator." In *Laser Congress 2021 (ASSL,LAC) (2021), Paper ATu2A.7*, ATu2A.7. Optica Publishing Group. doi:10.1364/ASSL.2021.ATu2A.7.

Wang, Y., J. Li, B. Zhai, Y. Hu, K. Mo, R. Lu, and Y. Liu. 2016. "Tunable and switchable dual-wavelength mode-locked $tm^{3+}$-doped fiber laser based on a fiber taper." *Optics Express* 24 (14): 15299–15306. doi:10.1364/OE.24.015299.

Wang, S., L. Liao, Y. Xing, H. Li, J. Peng, L. Yang, N. Dai, and J. Li. 2020a. "Promotion of pulse peak power by halving the repetition rate based on a vector soliton." *Optics Letters* 45 (7): 1635–1638. doi:10.1364/OL.384348.

Wang, Y., C. Wang, F. Zhang, J. Guo, C. Ma, W. Huang, Y. Song, Y. Ge, J. Liu, and H. Zhang. 2020b. "Recent advances in real-time spectrum measurement of soliton dynamics by dispersive fourier transformation." *Reports on Progress in Physics* 83 (11): 116401. doi:10.1088/1361-6633/abbcd7.

Wei, Y., B. Li, X. Wei, Y. Yu, and K. K. Y. Wong. 2018. "Ultrafast spectral dynamics of dual-color-soliton intracavity collision in a mode-locked fiber laser." *Applied Physics Letters* 112 (8): 081104. doi:10.1063/1.5020821.

Wong, J. H., K. Wu, H. H. Liu, C. Ouyang, H. Wang, S. Aditya, P. Shum, et al. 2011. "Vector solitons in a laser passively mode-locked by single-wall carbon nanotubes." *Optics Communications* 284 (7): 2007–2011. doi:10.1016/j.optcom.2010.12.048.

Wu, Z., S. Fu, K. Jiang, J. Song, H. Li, M. Tang, P. Shum, and D. Liu. 2016. "Switchable thulium-doped fiber laser from polarization rotation vector to scalar soliton." *Scientific Reports* 6 (1): 34844. doi:10.1038/srep34844.

Xu, Z. W., and Z. X. Zhang. 2013. "All-normal-dispersion multi-wavelength dissipative soliton Yb-doped fiber laser." *Laser Physics Letters* 10 (8): 085105. doi:10.1088/1612-2011/10/8/085105.

Xu, K., X. Zhao, Z. Wang, J. Chen, T. Li, Z. Zheng, and W. Ren. 2020. "Multipass-assisted dual-comb gas sensor for multi-species detection using a free-running fiber laser." *Applied Physics B* 126 (3): 39. doi:10.1007/s00340-020-7382-x.

Xu, Y., Y. Xu, Y. Xu, M. Erkintalo, M. Erkintalo, Y. Lin, S. Coen, et al. 2021. "Dual-microcomb generation in a synchronously driven waveguide ring resonator." *Optics Letters* 46 (23): 6002–6005. doi:10.1364/OL.443153.

Xu, K., L. Ma, J. Chen, X. Zhao, Q. Wang, R. Kan, Z. Zheng, and W. Ren. 2022b. "Dual-comb spectroscopy for laminar premixed flames with a free-running fiber laser." *Combustion Science and Technology* 194 (12): 2523–2538. doi:10.1080/00102202.2021.1879796.

Xu, S., A. Turnali, and M. Y. Sander. 2022a. "Group-velocity-locked vector solitons and dissipative solitons in a single fiber laser with net-anomalous dispersion." *Scientific Reports* 12 (1): 6841. doi:10.1038/s41598-022-10818-4.

Yan, M., Q. Hao, X. Shen, and H. Zeng. 2018. "Experimental study on polarization evolution locking in a stretched-pulse fiber laser." *Optics Express* 26 (13): 16086–16092. doi:10.1364/OE.26.016086.

Yang, Q.-F., Y. Xu, K. Y. Yang, and K. Vahala. 2017. "Counter-propagating solitons in micro-resonators." *Nature Photonics* 11 (9): 560–564. doi:10.1038/nphoton.2017.117.

Yao, X. 2014. "Generation of bidirectional stretched pulses in a nanotube-mode-locked fiber laser." *Applied Optics* 53 (1): 27–31. doi:10.1364/AO.53.000027.

Yun, L., and D. Han. 2012. "Evolution of dual-wavelength fiber laser from continuous wave to soliton pulses." *Optics Communications* 285 (24): 5406–5409. doi:10.1016/j.optcom.2012.07.114.

Yun, L., X. Liu, and D. Mao. 2012. "Observation of dual-wavelength dissipative solitons in a figure-eight erbium-doped fiber laser." *Optics Express* 20 (19): 20992–20997. doi:10.1364/OE.20.020992.

Zeng, C., X. Liu, and L. Yun. 2013. "Bidirectional fiber soliton laser mode-locked by single-wall carbon nanotubes." *Optics Express* 21 (16): 18937–18942. doi:10.1364/OE.21.018937.

Zeng, J., M. Y. Sander, M. Y. Sander, and M. Y. Sander. 2022. "Real-time observation of chaotic and periodic explosions in a mode-locked Tm-doped fiber laser." *Optics Express* 30 (5): 7894–7906. doi:10.1364/OE.449744.

Zhang, H., D. Y. Tang, L. M. Zhao, and N. Xiang. 2008. "Coherent energy exchange between components of a vector soliton in fiber lasers." *Optics Express* 16 (17): 12618. doi:10.1364/OE.16.012618.

Zhang, H., D. Y. Tang, X. Wu, and L. M. Zhao. 2009b. "Multi-wavelength dissipative soliton operation of an erbium-doped fiber laser." *Optics Express* 17 (15): 12692. doi:10.1364/OE.17.012692.

Zhang, H., D. Y. Tang, L. M. Zhao, X. Wu, and H. Y. Tam. 2009a. "Dissipative vector solitons in a dispersion-managed cavity fiber laser with net positive cavity dispersion." *Optics Express* 17 (2): 455–460. doi:10.1364/OE.17.000455.

Zhang, Z. X., Z. W. Xu, and L. Zhang. 2012. "Tunable and switchable dual-wavelength dissipative soliton generation in an all-normal-dispersion Yb-doped fiber laser with birefringence fiber filter." *Optics Express* 20 (24): 26736. doi:10.1364/OE.20.026736.

Zhao, L. M., D. Y. Tang, H. Zhang, and X. Wu. 2008b. "Polarization rotation locking of vector solitons in a fiber ring laser." *Optics Express* 16 (14): 10053–10058. doi:10.1364/OE.16.010053.

Zhao, L. M., D. Y. Tang, H. Zhang, X. Wu, and N. Xiang. 2008a. "Soliton trapping in fiber lasers." *Optics Express* 16 (13): 9528–9533. doi:10.1364/OE.16.009528.

Zhao, L. M., D. Y. Tang, X. Wu, and H. Zhang. 2010. "Dissipative soliton trapping in normal dispersion-fiber lasers." *Optics Letters* 35 (11): 1902–1904. doi:10.1364/OL.35.001902.

Zhao, X., Z. Zheng, Y. Liu, G. Hu, and J. Liu. 2014. "Dual-wavelength, bidirectional single-wall carbon nanotube mode-locked fiber laser." *IEEE Photonics Technology Letters* 26 (17): 1722–1725. doi:10.1109/LPT.2014.2332000.

Zhao, X., G. Hu, B. Zhao, C. Li, Y. Pan, Y. Liu, T. Yasui, and Z. Zheng. 2016. "Picometer-resolution dual-comb spectroscopy with a free-running fiber laser." *Optics Express* 24 (19): 21833–21845. doi:10.1364/OE.24.021833.

Zhao, X., T. Li, Y. Liu, Q. Li, and Z. Zheng. 2018. "Polarization-multiplexed, dual-comb all-fiber mode-locked laser." *Photonics Research* 6 (9): 853–857. doi:10.1364/PRJ.6.000853.

Zhao, K., H. Jia, P. Wang, J. Guo, X. Xiao, and C. Yang. 2019. "Free-running dual-comb fiber laser mode-locked by nonlinear multimode interference." *Optics Letters* 44 (17): 4323–4326. doi:10.1364/OL.44.004323.

Zhao, K., C. Gao, X. Xiao, and C. Yang. 2020a. "Buildup dynamics of asynchronous vector solitons in a polarization-multiplexed dual-comb fiber laser." *Optics Letters* 45 (14): 4040–4043. doi:10.1364/OL.398323.

Zhao, K., Y. Li, X. Xiao, and C. Yang. 2020b. "Nonlinear multimode interference-based dual-color mode-locked fiber laser." *Optics Letters* 45 (7): 1655–1658. doi:10.1364/OL.388314.

Zhao, K., C. Gao, X. Xiao, and C. Yang. 2021. "Real-time collision dynamics of vector solitons in a fiber laser." *Photonics Research* 9 (3): 289–298. doi:10.1364/PRJ.413855.

Zhu, S. N., Z. C. Wu, S. N. Fu, and L. M. Zhao. 2018. "Manipulation of group-velocity-locked vector dissipative solitons and properties of the generated high-order vector soliton structure." *Applied Optics* 57 (9): 2064. doi:10.1364/AO.57.002064.

Zhu, Y., F. Xiang, L. Jin, S. Y. Set, and S. Yamashita. 2019. "All-fiber dual-wavelength mode-locked laser using a bend-induced-birefringence Lyot-filter as gain-tilt equalizer." *IEEE Photonics Journal* 11 (6): 1–7. doi:10.1109/JPHOT.2019.2946380.

Zhu, Y., Z. Cui, X. Sun, T. Shirahata, L. Jin, S. Yamashita, and S. Y. Set. 2020. "Fiber-based dynamically tunable lyot filter for dual-wavelength and tunable single-wavelength mode-locking of fiber lasers." *Optics Express* 28 (19): 27250. doi:10.1364/OE.402173.

# 5 Vector Solitons Formed in Linearly Birefringent Single Mode Fibers

*X. Hu, J. Guo and D. Y. Tang*

## CONTENTS

5.1   Introduction ....................................................................................................144
5.2   Theoretical Studies ........................................................................................144
    5.2.1   Pulse Propagation in Linearly Birefringent SMFs ...........................144
    5.2.2   Vector Solitons Formed under Coherent XPC ..................................145
        5.2.2.1   Coherently Coupled Vector Bright Solitons .......................145
        5.2.2.2   Coherently Coupled Vector Black Solitons .......................147
        5.2.2.3   Coherently Coupled Vector Black-White Solitons .............148
    5.2.3   Vector Solitons Formed under Incoherent XPC ................................149
        5.2.3.1   Incoherently Coupled Vector Bright Solitons ...................149
        5.2.3.2   Incoherently Coupled Vector Gray Solitons .......................150
        5.2.3.3   Incoherently Coupled Vector Dark-Bright Solitons ...........151
5.3   Experimental Studies .....................................................................................153
    5.3.1   Experimental Setup .........................................................................153
    5.3.2   Coherently Coupled Vector Solitons ................................................155
        5.3.2.1   Vector Bright Solitons .........................................................155
        5.3.2.2   Vector Black Solitons .........................................................156
        5.3.2.3   Vector Black-White Solitons ..............................................157
    5.3.3   Incoherently Coupled Vector Solitons .............................................159
        5.3.3.1   Group Velocity Locked Bright Solitons ..............................159
        5.3.3.2   Incoherently Coupled Gray Solitons ..................................160
        5.3.3.3   Incoherently Coupled Dark-Bright Solitons ......................161
5.4   Outlook on Dissipative Vector Solitons ..........................................................164
5.5   Conclusions ...................................................................................................165
References .............................................................................................................165

This chapter discusses the different types of vector solitons formed in linearly birefringent SMFs. It begins with a brief introduction, followed by descriptions on the various forms of vector solitons theoretically predicted for the coupled NLSEs. Section 5.2 describes the experimental observations of vector solitons formed in SMFs. Finally, Sections 5.3 and 5.4 summarizes the research results and discusses on the possible future research directions.

DOI: 10.1201/9781003206767-5

## 5.1  INTRODUCTION

Solitons are localized nonlinear waves that can propagate over long distances in dispersive media while maintaining their shapes. Over decades, solitons have been intensively investigated in different fields such as hydrodynamics [1, 2], plasma physics [3], Bose-Einstein condensates [4], biophysics [5], and nonlinear optics [6, 7]. The soliton formation in single mode fibers (SMFs) has attracted great attention owing to its significance to the fundamental study and potential applications in long-haul optical communication systems and optical signal processing. The soliton formation in SMFs is mathematically described by the NLSE, a paradigm equation governs the nonlinear-wave propagation in many physical systems [8]. Either scalar bright or dark solitons can be formed in a SMF depending on the sign of fiber dispersion. Specifically, in the anomalous dispersion region the scalar bright solitons are formed, while in the normal dispersion region the scalar dark solitons are formed [9]. Experimentally, both the scalar bright and dark soliton formations in SMFs have been demonstrated [10, 11].

In practice, a SMF always possesses weak birefringence due to fiber bending and/ or imperfections. Therefore, a SMF supports two orthogonal polarization modes. Depending on the strength of birefringence, the two orthogonally polarized modes can couple with each other either incoherently or coherently. While the incoherent cross-polarization coupling process does not cause energy exchange between them, the coherent coupling is a kind of four-wave mixing (FWM) process that involves strong energy transfer between the coupling components. Nonlinear light pulse propagation in a linearly birefringent SMF is governed by the coupled NLSEs [12]. Comparing with the scalar NLSE, the cross-polarization coupling of light in a SMF could give rise to many novel types of vector solitons, such as the vector bright solitons [13], vector black solitons [14], vector dark-bright solitons [15, 16] and the polarization rotation vector solitons [17, 18]. This chapter gives a comprehensive study of the different types of vector solitons formed in linearly birefringent SMFs.

## 5.2  THEORETICAL STUDIES

### 5.2.1  PULSE PROPAGATION IN LINEARLY BIREFRINGENT SMFs

When an optical pulse propagates in a linearly birefringent fiber, it undergoes complicated changes due to the mutual interaction between the fiber dispersion and nonlinearity. Dispersion is an intrinsic feature of materials. It is defined as the variation of group velocity with frequency. In an SMF, there are two types of dispersion that could broaden the pulse widths: The chromatic dispersion and polarization mode dispersion. In parallel with the fiber dispersion effect, the fiber nonlinearity is referred to as the variation of the phase velocity as a function of pulse intensity. In optical fibers, the optical Kerr effect is one of the most common nonlinear effects. As there exist two orthogonal polarization components in a linearly birefringent SMF, the Kerr effect will not only generate self-phase modulation (SPM) on each component, but also generate cross-phase modulation (XPM) between them. Mathematically, nonlinear pulse propagation in a linear birefringent SMF is described by the following coupled NLSEs [12]:

$$\frac{\partial u}{\partial z} = i\beta u - \delta \frac{\partial u}{\partial t} - i\frac{\beta_2}{2}\frac{\partial^2 u}{\partial t^2} + i\gamma\left(|u|^2 + \frac{2}{3}|v|^2\right)u + i\frac{\gamma}{3}v^2 u^*$$

$$\frac{\partial v}{\partial z} = -i\beta v + \delta \frac{\partial v}{\partial t} - i\frac{\beta_2}{2}\frac{\partial^2 v}{\partial t^2} + i\gamma\left(|v|^2 + \frac{2}{3}|u|^2\right)v + i\frac{\gamma}{3}u^2 v^*$$

$$(5.1)$$

Where: $u$ and $v$ are the normalized slowly varying pulse envelopes along the slow and fast axes, $u^*$ and $v^*$ are the conjugates for $u$ and $v$; $2\beta = 2\Delta n/\lambda$ is the wave number difference between the two polarization modes; $\Delta n$ represents the refractive index difference between the two polarization axes; $2\delta = 2\beta\lambda/2\pi c$ is the group velocity mismatch, $\beta_2$ is the second-order dispersion coefficient; and $\gamma$ is the nonlinear coefficient of the fiber. The equations are written in a coordinate system that moves with the averaged group velocity.

Depending on the strength of fiber birefringence, the two polarization modes could be coupled either coherently or incoherently. Coherent cross-polarization coupling (XPC) only occurs in SMFs with very weak birefringence where the phase-matching condition between the two orthogonal polarization modes is automatically fulfilled. On the other hand, incoherent XPC occurs when the relative phase between the two modes varies much faster than the relaxation time of the nonlinear medium. It arises when the refractive index of one beam is modulated by another one. This process is mainly determined by the relative strength of the individual light beam and does not require the phase-matching condition.

## 5.2.2 Vector Solitons Formed under Coherent XPC

When the linear birefringence of a SMF is sufficiently small, the group velocity mismatch term in Equation (5.1) could be neglected. Writing in the dimensionless form, Equation (5.1) becomes the following:

$$i\frac{\partial U}{\partial \xi} - \frac{\eta}{2}\frac{\partial^2 U}{\partial \tau^2} + \left(|U|^2 + \frac{2}{3}|V|^2\right)U + \frac{1}{3}V^2 U^* + \kappa U = 0$$

$$i\frac{\partial V}{\partial \xi} - \frac{\eta}{2}\frac{\partial^2 V}{\partial \tau^2} + \left(|V|^2 + \frac{2}{3}|U|^2\right)V + \frac{1}{3}U^2 V^* - \kappa V = 0$$

$$(5.2)$$

Where $U = \sqrt{\gamma L_d}u, V = \sqrt{\gamma L_d}v, \xi = \dfrac{z}{L_d}, \tau = \dfrac{t}{T_0}$ and $\kappa = \dfrac{T_0^2}{2|\beta_2|}(\beta_{0x} - \beta_{0y})$. $T_0$ is a reference pulse width, $L_d = \dfrac{T_0^2}{|\beta_2|}$ is the dispersion length of the reference pulse, $\eta = \pm 1$. In the anomalous dispersion regime, $\eta = -1$, while in the normal dispersion regime, $\eta = 1$.

### 5.2.2.1 Coherently Coupled Vector Bright Solitons

We first look for vector bright solitary wave solutions of Equation (5.2) in the anomalous dispersion regime ($\eta = -1$). Assume that the two polarization components have the same propagation constant $\beta$ and are phase-locked at $0°$

$$U = u(\tau)\exp(-i\beta z)$$
$$V = v(\tau)\exp(-i\beta z)$$

(5.3)

Substitute Equation (5.3) into Equation (5.2), and following the potential method discussed in reference [14], we obtain the following approximate analytical solutions for the vector bright solitons:

$$\begin{cases} u(\tau) = A \sec h(B\tau) \\ v(\tau) = C \sec h(D\tau) \end{cases} \text{and} \begin{cases} D^2 = -2(\beta + \kappa) \\ B^2 = -2(\beta - \kappa) \\ A^2 + C^2 = -2\beta \end{cases}$$

(5.4)

Where $A, B, C$ satisfy $A^2 + C^2 = -2\beta = B^2$. Hence, the final expressions are given as,

$$\begin{cases} u(\tau) = \sqrt{-2\beta} \cos\theta \sec h\left(\sqrt{-2(\beta - \kappa)}\tau\right) \\ v(\tau) = \sqrt{-2\beta} \sin\theta \sec h\left(\sqrt{-2(\beta + \kappa)}\tau\right) \end{cases}$$

(5.5)

Where $\theta$ is an arbitrary angle variable. When $k = 0$, the two polarization components would have equal pulse width, and Equation (5.5) give the exact solution of Equation (5.2). A typical coherently coupled vector bright soliton is shown in Figure 5.1.

The vector bright solitons are also known as the phase-locked bright-bright solitons. It is worth mentioning that the phase-locked bright-bright solitons can only appear in optical fibers with anomalous dispersion under sufficiently small birefringence. At a fixed dispersion, as described by Equation (5.4), when the fiber birefringence is increased, the pulse width of the solitons along one polarization direction narrows down, while that along the orthogonal direction broadens.

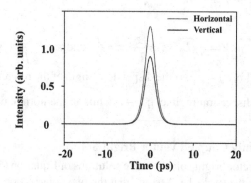

**FIGURE 5.1**   Coherently coupled bright solitons with $\beta = -1$, $k = -0.1$.

### 5.2.2.2 Coherently Coupled Vector Black Solitons

We next look for vector black solitary wave solutions of Equation (5.2) in the normal dispersion regime ($\eta = 1$). Assuming the two polarization components have the same propagation constant $\beta < 0$ but are phase-locked at $\pi/2°$

$$U = u(\tau)\exp(-i\beta z)$$
$$V = iv(\tau)\exp(-i\beta z) \tag{5.6}$$

Substitute Equation (5.6) into Equation (5.2), and following the potential method analysis discussed in reference [14], we obtained the following approximate vector black soliton solutions:

$$\begin{cases} u = A\tan h(D\tau) \\ v = C\tan h(E\tau) \end{cases} \text{ and } \begin{cases} A = \sqrt{\dfrac{-3\beta}{4} + \dfrac{-3\kappa}{2}} \\[2mm] C = \sqrt{\dfrac{-3\beta}{4} + \dfrac{3\kappa}{2}} \\[2mm] D = \sqrt{-\beta - k} \\[2mm] E = \sqrt{-\beta + k} \end{cases} \tag{5.7}$$

Where $A, C, D, E$ satisfy $\begin{cases} A^2 - D^2 + \dfrac{1}{3}C^2 = 0 \\[2mm] C^2 - E^2 + \dfrac{1}{3}A^2 = 0 \end{cases}$. The approximate solutions are

valid for small $k$ and become the accurate analytical solution of Equation (5.2) when $k = 0$. A typical black-black vector soliton is shown in Figure 5.2.

The vector black solitons have the following unique characteristics: (a) The intensity dips of the solitons always drop to zero; (b) At the centre of the intensity dip, there is always a $\pi$ phase jump; (c) The pulse widths of the solitons are determined by the CW background intensity. The stronger the CW is, the narrower the pulse

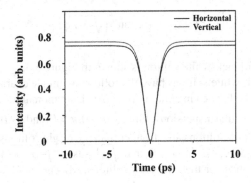

**FIGURE 5.2** Coherently coupled dark solitons with $\beta = -1$, $k = -0.01$.

width; (d) The vector black solitons are highly robust and exist around $\kappa = 0$; (e) At a fixed CW intensity, as the fiber birefringence is increased, the pulse width of the soliton along one polarization direction narrows down, while that along the orthogonal direction broadens.

### 5.2.2.3   Coherently Coupled Vector Black-White Solitons

The above results show that under coherent coupling, vector bright solitons could be formed in the anomalous dispersion regime while vector black solitons could be formed in the normal dispersion regime. These vector solitons have the characteristic that each of their polarization components could exist as a scalar soliton, even when there is no coupling between them. We next look for vector black-white solitary wave solutions whose two polarization components have the same propagation constant $\beta$ and are phase-locked at $0°$

$$U = u(\tau)\exp(-i\beta z)$$
$$V = v(\tau)\exp(-i\beta z) \tag{5.8}$$

Setting Equation (5.8) into Equation (5.2), and following the potential method analysis shown in reference [14], we obtained the following vector black-white soliton solutions,

$$u(\tau) = A \sec h(B\tau) \quad \text{and} \quad \begin{cases} A^2 = -\beta - 3\kappa \\ B^2 = 4\kappa \\ C^2 = -\beta + \kappa \end{cases} \tag{5.9}$$

Where $A, B, C$ satisfy $\begin{cases} A^2 = -B^2 + C^2 \\ A^2 + \dfrac{1}{2}B^2 = -(\beta + \kappa) \end{cases}$. Therefore, the final soliton solutions are given by,

$$\begin{cases} u(\tau) = \sqrt{-\beta - 3\kappa} \sec h\left(\sqrt{4\kappa/\eta}\tau\right) \\ v(\tau) = \sqrt{-\beta + \kappa} \tan h\left(\sqrt{4\kappa/\eta}\tau\right) \end{cases} \tag{5.10}$$

A typical vector black-white soliton is shown in Figure 5.3.

The vector black-white solitons have the following unique characteristics: (a) The vector solitons can be formed in either normal ($\eta = 1$) or anomalous dispersion regime ($\eta = -1$); (b) In the normal dispersion regime, $-\beta > 0, \kappa > 0$, and $\sqrt{-\beta - 3\kappa} < \sqrt{-\beta + \kappa}$, indicating that the dark solitons are stronger than the bright solitons; (c) In the anomalous dispersion regime, $-\beta > 0, \kappa < 0$, and $\sqrt{-\beta - 3\kappa} > \sqrt{-\beta + \kappa}$, indicating that the bright solitons are stronger than the dark solitons; (d) The pulse widths of the dark and bright solitons are always the same. Under fixed group-velocity dispersion, they are uniquely determined by the birefringence of the fibers [15].

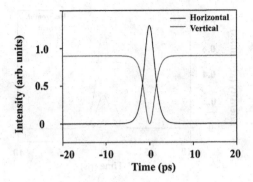

**FIGURE 5.3**    Coherently coupled black-white solitons with $\beta = -1$, $k = -0.1$.

### 5.2.3 Vector Solitons Formed under Incoherent XPC

When the linear birefringence of a SMF is large, the four-wave mixing (FWM) term in Equation (5.1) can be neglected. Consequently, the coupling between the two polarization components becomes incoherent. Written in the dimensionless form, Equation (5.1) can be written in the following forms:

$$i\frac{\partial U}{\partial \xi} + \delta\frac{\partial U}{\partial \tau} - \frac{\eta}{2}\frac{\partial^2 U}{\partial \tau^2} + \left(|U|^2 + \frac{2}{3}|V|^2\right)U = 0$$

$$i\frac{\partial V}{\partial \xi} - \delta\frac{\partial U}{\partial \tau} - \frac{\eta}{2}\frac{\partial^2 V}{\partial \tau^2} + \left(|V|^2 + \frac{2}{3}|U|^2\right)V = 0$$

$$(5.11)$$

Where $U$, $V$, $\xi$, $\tau$ have the same definition as shown for Equation (5.2), and $\delta = \frac{T_0}{2|\beta_2|}\left(\beta_{1x} - \beta_{1y}\right)$ is the normalized group velocity mismatch between the two polarization modes.

#### 5.2.3.1 Incoherently Coupled Vector Bright Solitons

Menyuk first numerically studied Equation (5.11) using the following initial conditions [19]:

$$u(0,\tau) = A\cos\theta \sec h(\tau)$$
$$v(0,\tau) = A\sin\theta \sec h(\tau)$$

$$(5.12)$$

Where $A$ is the soliton amplitude and $\theta$ is the angle of the soliton polarization direction with respect to the horizontal axis of the fiber. He found that when the soliton amplitude is weak, due to the fiber birefringence, the two solitons polarized along the two orthogonal polarization modes of the fiber will separate from each other. However, when it is larger than a certain threshold, which depends nonlinearly on the strength of the linear birefringence, the two solitons can shift their frequency in opposite directions to compensate the fiber birefringence and propagate at the same group velocity. A typical such formed vector bright soliton is presented in Figure 5.4.

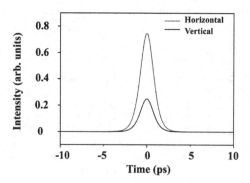

**FIGURE 5.4** Incoherently coupled bright solitons with $N = 1$, $\theta = \dfrac{\pi}{4}$.

 

    Incoherently coupled bright solitons are also known as the group velocity locked bright solitons. Further theoretical studies have shown that, as a result of the incoherent XPC, the bright solitons with slightly different group velocities can trap each other through dynamically shifting their central wavelengths [20, 21]. However, if the birefringence of a fiber becomes sufficiently strong, the two polarization components evolve independently and separate from each other because of the group velocity mismatch.

### 5.2.3.2 Incoherently Coupled Vector Gray Solitons

Not only incoherently coupled bright solitons but also incoherently coupled grey solitons could be trapped together and propagate as a unit. Kivshar and Turitsyn have first theoretically studied the interaction of two optical fields in the normal dispersion SMFs [22] and predicted the formation of incoherently coupled vector gray solitons. They showed that vector dark solutions do exist in the form of two partial dark solitons excited on different CW backgrounds in the form,

$$U = U_0 \left( \cos\phi_1 \tan hZ + i \sin\phi_1 \right) \exp\left( i\Theta_1 \right)$$
$$V = V_0 \left( \cos\phi_2 \tan hZ + i \sin\phi_2 \right) \exp\left( i\Theta_2 \right) \tag{5.13}$$

Where $Z = v\left( \tau + \xi/W - \tau_0 \right)$, and,

$$\Theta_1 = k_1\tau + \left( U_0^{\,2} + \sigma V_0^{\,2} + k_1^{\,2} \right)\xi$$
$$\Theta_2 = k_2\tau + \left( V_0^{\,2} + \sigma U_0^{\,2} + k_2^{\,2} \right)\xi \tag{5.14}$$

And the parameters $U_0, V_0, v, W, k_1,$ and $k_2$ are coupled by the following relations,

$$U_0 \cos\phi_1 = V_0 \cos\phi_2$$

$$v^2 = \frac{1+\sigma}{2} U_0^2 \cos^2\phi_1$$

$$W^{-1} = U_0 \sqrt{\frac{1+\sigma}{2}} \frac{\sin(\phi_1+\phi_2)}{\cos\phi_2} + (k_1+k_2)$$

$$k_2 - k_1 = U_0 \sqrt{\frac{1+\sigma}{2}} \frac{\sin(\phi_1-\phi_2)}{\cos\phi_2} \tag{5.15}$$

The pulse intensities in each mode may be calculated to be,

$$|U|^2 = U_0^2 \left(1 - \frac{\cos^2\phi_1}{\cos h^2 Z}\right)$$

$$|V|^2 = V_0^2 \left(1 - \frac{\cos^2\phi_2}{\cos h^2 Z}\right) \tag{5.16}$$

Based on Equation (5.16), a vector gray soliton has the form shown in Figure 5.5.

The vector gray solitons have the following characteristics: (a) The intensity dips of the gray solitons do not drop to zero; (b) At the center of the intensity dips, the phase jump is always less than $\pi$; (c) Once different vector gray solitons are formed, they travel at different group velocities and undergo pulse splitting during propagation.

### 5.2.3.3   Incoherently Coupled Vector Dark-Bright Solitons

M. Lisak et al. were the first to offer a theoretical study of the incoherently coupled vector dark-bright soliton formation [16]. Generally, the vector dark-bright solitons should satisfy the following incoherently coupled NLSEs:

$$i\frac{\partial u_b}{\partial z} - \beta_b \frac{\partial^2 u_b}{\partial t^2} + \gamma\left(|u_b|^2 + \alpha|u_d|^2\right)u_b = 0.$$

$$i\frac{\partial u_d}{\partial z} - \beta_d \frac{\partial^2 u_d}{\partial t^2} + \gamma\left(|u_d|^2 + \alpha|u_b|^2\right)u_d = 0. \tag{5.17}$$

Where $u_b$ and $u_d$ are the normalized optical fields of the bright and dark solitons, $\beta_b$ and $\beta_d$ are their group-velocity-dispersion (GVD) coefficients, $\gamma$ is the fiber's nonlinearity coefficient, and $\alpha$ is the cross-coupling coefficient. For the case of XPC, $\alpha = 2/3$. It was shown by Lisak et al., Equation (5.17) admit the following coupled dark-bright (DB) soliton solutions,

$$u_b = A_b \sec h\left(B\xi\right)\exp\left[i\left(\sigma_b z - \Omega_b t\right)\right].$$

$$u_d = A_d \left(1 - C^2 \sec h^2 B\xi\right)^{\frac{1}{2}} \exp\left[i\left(\sigma_d z - \Omega_d t + \varphi_d\right)\right]. \tag{5.18}$$

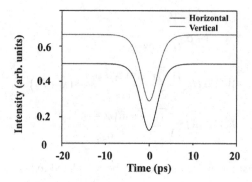

**FIGURE 5.5**  Incoherently coupled gray solitons with $U_0 = V_0 = 0.1$, $\phi_1 = \dfrac{\pi}{6}$.

Where $\xi = t - \dfrac{z}{v}$, $\dfrac{1}{v} = 2\beta_b\Omega_b = 2\beta_d\Omega_d$, $A_b$, $A_d$ and $v$ are arbitrary constants, and

$$B^2 = \frac{1-\alpha^2}{2} \frac{\gamma}{\alpha\beta_d - \beta_b} A_b^2. \tag{5.19}$$

$$\sigma_b = \gamma\left[\alpha A_d^2 - \frac{1-\alpha^2}{2} \frac{\beta_b}{\alpha\beta_d - \beta_b} A_b^2\right] + \frac{1}{4\beta_b v^2}. \tag{5.20}$$

$$\sigma_d = \gamma\left[\frac{\alpha\beta_b + \left[(\alpha^2-3)/2\right]\beta_d}{\alpha\beta_b - \beta_d} A_d^2 - \frac{1-\alpha^2}{2} \frac{\beta_d}{\alpha\beta_d - \beta_b} A_b^2\right] + \frac{1}{4\beta_d v^2}. \tag{5.21}$$

$$\varphi_d = \frac{N}{2\beta_d A_d^2} \int^{\xi} \frac{d\xi}{\left(1 - C^2 \sec h^2 B\xi\right)^{\frac{1}{2}}}. \tag{5.22}$$

$$N^2 = \frac{2\beta_d^2 A_d^6 \gamma}{\beta_d - \alpha\beta_b}\left(1-\alpha^2\right)\left(2 - C^2\right). \tag{5.23}$$

$$C^2 = \frac{\alpha\beta_b - \beta_d}{\beta_b - \alpha\beta_d} \frac{A_b^2}{A_d^2}. \tag{5.24}$$

The vector dark-bright soliton solutions are possible only for certain sign combinations of the group velocity dispersion:

i.  $\beta_b < 0, \beta_d > 0$, solutions are always possible.
ii. $\beta_b < 0, \beta_d < 0$, solutions are possible only if $\beta_d < \alpha\beta_b$.

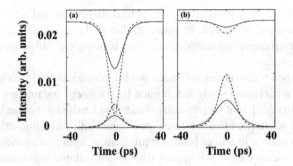

**FIGURE 5.6** Typical intensity profiles of the theoretically predicted coupled dark-bright solitons. (a) In the normal-GVD fiber, with $\beta_b = 0.5\,\text{ps}^2/\text{km}$, $\beta_b = 1\,\text{ps}^2/\text{km}$, $A_d^2 = 22.5\,\text{mW}$, $A_b^2 = 2.5\,\text{mW}$ (solid lines), and $A_b^2 = 4.9\,\text{mW}$ (dashed lines). (b) In the anomalous-GVD fiber, with $\beta_b = -1\,\text{ps}^2/\text{km}$, $\beta_d = -0.5\,\text{ps}^2/\text{km}$, $A_d^2 = 10\,\text{mW}$, $A_b^2 = 2.5\,\text{mW}$ (solid lines), and $A_b^2 = 4.9\,\text{mW}$ (dashed lines). Reprinted with permission from Hu et al. (2019) [37] © The Optical Society.

iii.  $\beta_b > 0, \beta_d > 0$, solutions are possible only if $\beta_b < \alpha\beta_d$.

iv.  $\beta_b > 0, \beta_d < 0$, no solution is possible.

It is worth mentioning that the abovementioned relations are only valid when $\alpha < 1$, if $\alpha \geq 1$, all the relations reverses. We consider the case when the vector dark-bright solitons are formed within the same dispersion regime. The Intensity profiles of two such vector dark-bright solitons are plotted in Figure 5.6. Different from their scalar counterparts, the coupled dark-bright solitons can be formed in fibers with either the normal or the anomalous GVD; however, in different dispersion regimes the GVD coefficients $\beta_d$ and $\beta_b$ of the dark and bright soliton components should satisfy different conditions. While in the normal-GVD fibers the condition is (2/3) $\beta_d > \beta_b$, in the anomalous GVD fibers it is (2/3) $\beta_d < \beta_b$. The coupled bright and dark solitons have the same pulse width, which is mainly determined by the strength of the bright soliton component. The darkness of the dark soliton is determined by both the background light intensity and the strength of the bright soliton. If multiple dark-bright solitons are formed in the same fiber, they could have different pulse heights and widths.

## 5.3  EXPERIMENTAL STUDIES

### 5.3.1  EXPERIMENTAL SETUP

Although various types of vector solitons were theoretically predicted for the non-linear light propagation in linearly birefringent SMFs, and in practice a SMF always exhibits birefringence, except the formation of the group velocity locked bright solitons [23], the experimental observation of other types of the theoretically predicted vector solitons are difficult to be realized. A major challenge for the experimental realization of the vector solitons is that their formation demands stringent conditions which could not be easily satisfied, e.g., in order to form the phase-locked vector bright solitons, SMFs with uniform anomalous dispersion and weak birefringence over a long distance is needed. In practice, due to the sensitive dependence of the linear fiber birefringence on temperature, fiber bending etc., this condition could not be easily achieved.

It is well known that under the mean field approximation, the dynamics of light circulation in a cavity is mainly determined by the cavity parameters averaged over the whole cavity [24], and theoretically it can also be shown that under steady-state operation and when the effective laser gain bandwidth limitation effect is neglectable, light circulation in a fiber laser is equivalent to light propagation in an endless fiber [25]. Therefore, it is anticipated that under suitable conditions, the various NLSE type of vector solitons could also be formed in a fiber laser. Moreover, as the effective cavity dispersion and birefringence of a fiber laser can be deliberately controlled by using the cavity dispersion- and birefringence-management techniques, there is a good chance that these theoretically predicted vector solitons could be experimentally revealed in fiber lasers. We note that the first phase-locked vector bright soliton was indeed experimentally demonstrated by Cundiff et al. in a mode-locked fiber laser [26].

Therefore, we have experimentally investigated the NLSE type of vector soliton formation in SMF lasers. The fiber lasers we used have a cavity configuration as schematic shown in Figure 5.7. They all have a ring cavity consisting of 3 m erbium-doped fiber (EDF) with a group velocity dispersion (GVD) coefficient of $\beta_2 = 63.4 \text{ ps}^2/\text{km}$, different lengths of SMF with a GVD coefficient of $\beta_2 = -23.8 \text{ ps}^2/\text{km}$ and dispersion compensating fiber (DCF) with a GVD coefficient of $\beta_2 = 5.1 \text{ ps}^2/\text{km}$, respectively. Both the SMF and DCF fibers are used because through the selection of appropriate DCF and SMF lengths, not only the desired averaged cavity GVD but also a desired cavity length can be achieved. The fiber laser is pumped by a 1,480 nm single mode Raman fiber laser which can provide a maximum output power of about 5 W. A polarization-independent isolator is inserted into the cavity to force the unidirectional circulation of light in the cavity. An intracavity polarization controller (PC) is used to fine-tune the linear cavity birefringence. A wavelength division multiplexer (WDM) is used to couple the pumping light into the cavity, and a 10% fiber output coupler is used to output the light. To separate the two orthogonal polarization components of the laser emission, the laser output is first sent

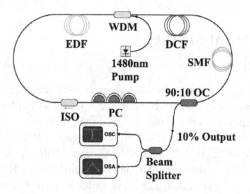

**FIGURE 5.7**   Schematic of experimental setup.

to a fiber pigtailed polarization beam splitter and then monitored with a high-speed electronic detection system consisting of 40 GHz photodetectors and a 33 GHz bandwidth real-time oscilloscope. A polarization controller is inserted between the laser output and the polarization beam splitter to balance the linear polarization change caused by the lead fibers. An optical spectrum analyser is used in our experiments to monitor the optical spectrum of the laser emission, and a commercial autocorrelator is used to measure the actual pulse width of the formed solitons.

### 5.3.2 Coherently Coupled Vector Solitons

#### 5.3.2.1 Vector Bright Solitons

We note that, different from the fiber laser used by Cundiff et al. [26], there is no saturable absorber in the laser cavity. Therefore, no mode-locking could automatically occur in the lasers. Generally, under low pump power our lasers always emit continuous waves (CW). However, depending on the net cavity birefringence, the laser emission could exhibit different features. A most frequent situation is that the laser simultaneously emits CW along the two orthogonal polarization directions of the cavity with different wavelengths, respectively. Carefully tuning the orientation of the intra-cavity PC, which alters the net linear cavity birefringence, the oscillation wavelength difference between the two orthogonally polarized CW emissions could be tuned. In our experiments we used the wavelength separation as an indicator on the strength of the net cavity birefringence and operated the lasers under different pump strengths and net cavity birefringence.

At very small net cavity birefringence and a pump power of 2 W, when the intra-cavity lasing intensity is about 630 mW, by carefully setting the orientation of the intra-cavity PC, the CW emission along the two orthogonal polarization directions could suddenly break up into a periodic bright pulse train, as shown in Figure 5.8a. The vector bright pulses are always periodically distributed in the cavity and display the same soliton pulse width and energies. Figure 5.8b presents the corresponding polarization resolved optical spectra. The spectra of the light along both polarization directions are significantly broadened; for example, the 3-dB bandwidth is estimated as 8 nm. Further, the Kelly sidebands are formed on the spectrum. The appearance of Kelly sidebands on the pulse spectra unambiguously shows that they are solitons [27]. The central wavelength of the solitons formed along the two orthogonal polarization directions are almost overlapped at 1,578 nm. The Kelly sidebands along the two polarization directions are also located at the same wavelength. If the spectrum of the total laser emission is measured, only one set of Kelly sidebands is observed. Besides, the FWM energy exchange sidebands are also observable, indicating that the bright solitons are phase-locked with each other [28]. Figure 5.8c shows the autocorrelation trace of the bright solitons. If a sec $h^2$ shape is assumed, it has a FWHM width of about 1.2 ps. Starting from the state, if the intra-cavity PC paddles is slightly tuned, the net cavity birefringence would be increased. To a certain range of the net cavity birefringence, phase-locked vector solitons could still be obtained. However, as shown by the polarization resolved soliton spectra and the positions of the Kelly sidebands, it could be deduced that the formed vector bright solitons would have unequal pulse widths, which agrees well with the theoretical predictions.

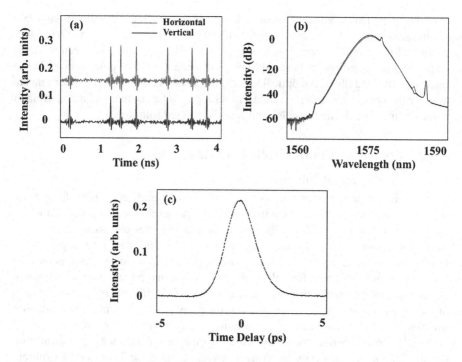

**FIGURE 5.8**  Phase-locked bright solitons. (a) Polarization resolved laser emission. (b) Optical spectrum. (c) Autocorrelation trace.

### 5.3.2.2  Vector Black Solitons

To observe vector black solitons, we changed the average cavity dispersion to the normal dispersion regime, by using 9 m single-mode fiber (SMF) and 15 m dispersion compensation fiber (DCF) to construct the laser cavity [29]. The averaged net cavity GVD coefficient is 1.99 ps$^2$/km. Initially, the fiber laser always emits CW along its two principal cavity polarization axes. By carefully tuning the intracavity PC paddles until the central wavelengths of the two orthogonally polarized CW laser emissions overlap with each other, we obtained the coherently coupled vector black solitons as shown in Figure 5.9a. Figure 5.9b shows the corresponding polarization resolved optical spectra. Multiple pairs of vector black solitons are simultaneously formed in the cavity. They are well scattered in the cavity, and no soliton bunching are observed. The vector black solitons are frozen in the cavity, no gray solitons are found to coexist with them. It is notable that the black solitons along the two orthogonal polarization directions have nearly the same spectral intensities and widths [28]. Based on the measured 3 dB spectral bandwidth of 0.12 nm, if a $\tan h^2$ pulse profile is assumed, the estimated pulse width of the black solitons is about 22 ps. At a pump power of 100 mw, the laser output power is measured as 0.8 mW, suggesting that the intracavity power is ~8 mW. Based on $P_0 = 3.11|\beta_2|/(\gamma\tau^2)$, the black solitons formed in our laser would have a pulse width of ~ 17 ps. Therefore, we believe the dark pulses are black solitons formed in the fiber laser. We also verified the properties of the black solitons under different pump intensities, as displayed in Figure. 5.9c.

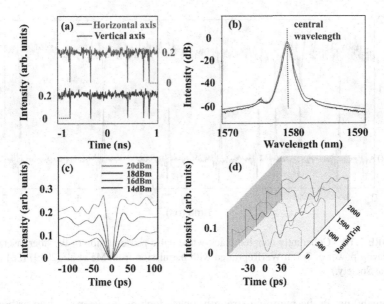

**FIGURE 5.9** (a) Coherently coupled vector black solitons. (b) Polarization resolved optical spectra. (c) Black soliton pulse width changes with pump intensity. (d) Evolution of black solitons. Reprinted with permission from Hu et al. (2020) [29] © The Optical Society.

Associated with the pump power increase, the CW level increases, and the pulse width of the black solitons narrows down, which agrees well with the results shown in Figure 5.2.

Continuously increasing the pump intensity to above 20 dBm, limited by the bandwidth of our detection system, the black solitons become no longer detectable. To check the stability of the vector black solitons, we have focused on one black soliton pulse and measured its evolution within thousands of cavity roundtrips. The result is displayed in Figure 5.9d. We confirmed that the formed black soliton remains stable at least within thousands of the cavity roundtrips. Further, the black soliton remains static in the cavity, and the relative position between black solitons within different cavity round trips remains unchanged.

### 5.3.2.3   Vector Black-White Solitons

In addition of the phase-locked dark-dark and bright-bright solitons, they are formed in either the normal or the anomalous fiber dispersion regime, phase-locked dark-bright vector solitons are also obtained under the resonant polarization coupling [30]. Using a fiber ring laser with an averaged cavity dispersion coefficient of ~2.1 ps²/km (3 m EDF, 13.4 m standard SMF (SMF28) 12.9 m dispersion-shifted fiber), a laser operation state, as shown in Figure 5.10, is observed. The two traces shown are the polarization-resolved laser emissions. It clearly demonstrates that while along one polarization direction the laser emitted a bright pulse, along the orthogonal polarization direction it emitted simultaneously a black pulse. Many coupled black-white pulse pairs are formed in the cavity, and, in particular, all the pairs have almost identical pulse parameters.

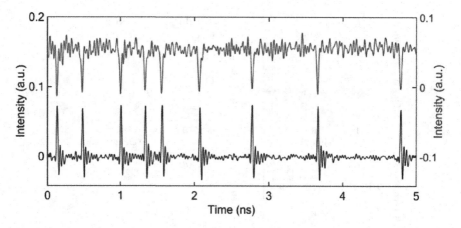

**FIGURE 5.10**  Coherently coupled black-white pulse pair emission of the fiber laser under intracavity power of 700 mW. (Reprinted with permission from Ma et al. (2019) [30] © The Optical Society.)

Depending on the pumping strength, phase locking between the two orthogonal polarization-modes could be achieved in a certain range of the net cavity birefringence, where the coupled dark-bright pulse pair emission was always obtained. However, depending on the net cavity birefringence, the soliton pulse width could change significantly. The pulse train shown in Figure 5.10 was obtained at a net cavity birefringence close to zero where the solitons have broad pulse width. Unfortunately, such a state was very sensitive to the net cavity birefringence. Slight turning the orientation of the intra-cavity PC to one direction, the state would suddenly terminate, while turning it in the opposite direction, the pulse width as well as the number of pulses in the cavity would quickly decrease, and eventually the dark pulses would become undetectable.

We experimentally characterized the features of the dark-bright soliton pairs at relatively large net cavity birefringence where the pulse width is insensitive to small net cavity birefringence change. Figure 5.11a shows again the polarization resolved emissions of the laser. Five pairs of the dark-bright pulses coexisted in the cavity, and they repeated with the cavity roundtrip time. Figure 5.11b is the corresponding optical spectra. Both the Kelly sidebands [27] and the coherent energy exchange spectral sidebands [28] are obviously observable in the spectra, which clearly shows that both the bright and dark pulses are solitons. Using a commercial autocorrelator, we measured the width of the bright pulses. It had a FWHM width of 1.53 ps. Assuming that the pulse had a $sech^2$ profile, its width was estimated ~990 fs, as shown in Figure 5.11c. Due to the low repetition rate of the pulses, the width of the dark solitons could not be measured with the autocorrelation method. However, the Kelly sidebands of the dark solitons had exactly the same positions as those of the bright solitons, indicating that they should have the same pulse width and properties.

The black-white vector solitons were also obtained in fiber lasers with other net cavity dispersions and cavity lengths. Through measuring the total laser output, we could also confirm that in the net normal cavity dispersion regime, the dark solitons have stronger pulse intensity than the bright solitons, while in the anomalous cavity

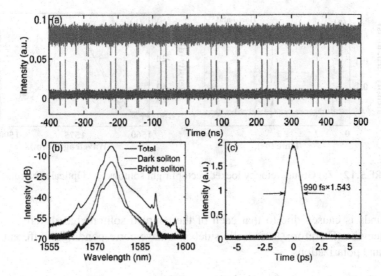

**FIGURE 5.11** Dark-bright pulse pair emission of the fiber laser measured at a relatively larger net cavity birefringence than the state shown in Figure 5.10. (a) Polarization resolved laser emissions. (b) Optical spectra. (c) Autocorrelation trace of the bright solitons. (Reprinted with permission from Ma et al. (2019) [30] © The Optical Society.)

dispersion regime, it is reversed. All features of the observed black-white vector solitons are found to be strongly in agreement with those of the coherently coupled dark-bright vector solitons theoretically predicted by Christodoulides [15].

### 5.3.3 INCOHERENTLY COUPLED VECTOR SOLITONS

#### 5.3.3.1 Group Velocity Locked Bright Solitons

Zhao et al. were the first to report the experimental observation of GVLVS in a passively mode-locked fiber laser [31]. With a semiconductor saturable absorber in the cavity, self-started mode-locking is achieved when the pump power is increased above the mode-locking threshold. When the net cavity birefringence is large, two sets of the mode locked pulses each polarized along one of the two principal cavity polarization axes are formed. The central wavelength and the moving speed of the solitons polarized along different polarization directions are different. If the oscilloscope is triggered with solitons polarized along one direction, all the solitons polarized along the same direction become stationary on the oscilloscope screen, while the solitons polarized along the orthogonal polarization direction still move on the screen, indicating that the solitons polarized along the two orthogonal polarization directions have different velocities. Carefully reducing the net cavity birefringence, all solitons in the cavity become moving at the same velocity.

Although in our fiber laser there is no saturable absorber in the cavity, the GVLVSs have also been obtained, as shown in Figure 5.12. One unique feature of GVLVSs is that the solitons have different central wavelengths but that they propagate at the same velocity in the cavity. On the total laser emission, two sets of Kelly sidebands, as shown in Figure 5.12a, could be obtained. The appearance of two sets of Kelly

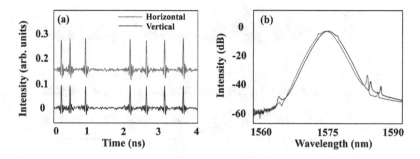

**FIGURE 5.12**    (a) Group-velocity locked vector bright solitons. (b) Optical spectrum.

sidebands is caused due to that each of the coupled solitons has different central wavelengths. Therefore, their Kelly sidebands have correspondingly different locations and polarizations.

### 5.3.3.2    Incoherently Coupled Gray Solitons

Scalar dark soliton formation in SMFs was experimentally studied by different research groups previously [11, 32, 33]. However, the formation of dark solitons in fiber lasers has been less addressed. Tang et al. firstly reported experimental evidence of scalar dark soliton formation in fiber lasers [34]. It was found that different from the multiple bright solitons formed in a fiber laser, which always have the same pulse energy, the dark solitons have random darkness and pulse widths. Based on a weak birefringent cavity fiber laser, Shao et al. further experimentally demonstrated vector gray soliton formation in a net normal dispersion fiber laser [35]

Shao et al. used a fiber laser made of 3 m EDF, 8.4 m SMF and 7 m DCF, and the average cavity GVD parameter is about 1.13 ps/km/nm. By operating the fiber laser in relatively large net cavity birefringence, at a pump power of 1.3 W, they observed that many gray solitons could be simultaneously formed in the cavity and they exhibit random pulse intensities, pulse widths, and pulse distributions. The features of the dark solitons formed along each of the two orthogonal polarization directions of the cavity resemble those of the scalar dark solitons reported [34]. However, as the net cavity birefringence is reduced, a dark soliton emission state, as shown in Figure 5.13, is further obtained. The gray solitons formed along the two orthogonal polarization directions are always synchronized. The temporal resolution of the detection system used is about 30 ps. Some of the dark pulses are limited by the resolution of the detection system. Figure 5.13b is the corresponding polarization-resolved optical spectra. The central wavelength of the solitons along each of the polarization directions is different, suggesting that they are incoherently coupled. It is worth mentioning that the spectral sidebands appeared in the spectrum are not the dark soliton Kelly sidebands but those formed due to the cavity induced modulation instability (CAMI) effect [36]. The experiments clearly shown that under incoherent XPC, orthogonally polarized dark solitons could couple together to form the group velocity-locked dark solitons. The formed vector dark solitons are stable even when the background intensity is periodically modulated. The experimental results agree strongly with the theoretical predictions and numerical simulations.

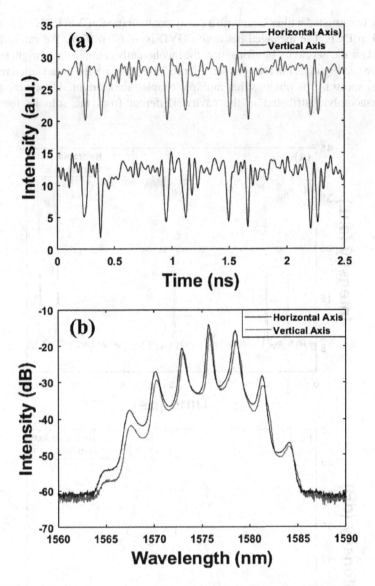

**FIGURE 5.13** Dark soliton emission of the fiber laser under weak cavity birefringence. (a) Polarization resolved oscilloscope traces. (b) Polarization resolved optical spectra. There is slight wavelength offset between the two spectra.

### 5.3.3.3  Incoherently Coupled Dark-Bright Solitons

Although the coupled dark-bright solitons have been theoretically predicted more than two decades ago [16], the experimental demonstration of such soliton complexes has been less studied. Operating a fiber laser with small net cavity dispersion and birefringence, which could be easily achieved using the cavity dispersion and birefringent management techniques, the coupled dark-bright solitons could be routinely generated [37].

We constructed a fiber laser with a cavity configuration of 3 m EDF, 10.5 m SMF, and 9 m DCF. The averaged net cavity GVD is −0.67 ps²/km. We carefully controlled for the net cavity birefringence, the incoherently coupled dark-bright solitons, as shown in Figure 5.14a are obtained. Figure 5.14b displays the polarization resolved optical spectra. It is obvious that multiple coupled dark-bright solitons are formed and randomly distributed in the cavity. Different from the solitons formed in

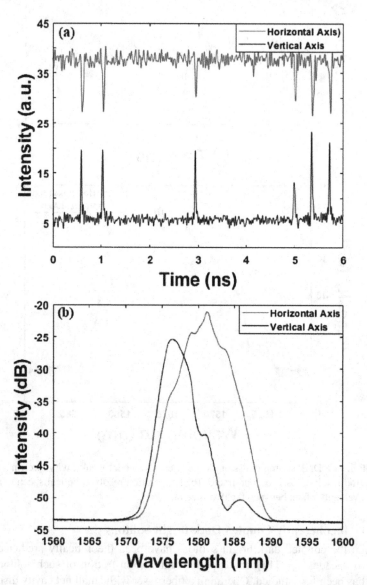

**FIGURE 5.14**    A typical picture of the emission of XPM-coupled dark-bright solitons, emitted by the fiber laser, in the net anomalous-cavity-GVD regime. (a) Oscilloscope traces of the polarization-resolved laser emission. (b) The corresponding optical spectra. Reprinted with permission from Hu et al. (2019) [37] © The Optical Society.

mode-locked fiber lasers, which always display the same soliton pulse width and heights, different pairs of dark-bright solitons feature different soliton pulse heights and widths. Based on the measured optical spectra of the polarization-resolved solitons, they obviously are incoherently coupled. We also checked the formation of dark-bright solitons in a fiber laser with a net normal cavity GVD of 0.67 ps$^2$/km, by cutting away a piece (0.22 m) of SMF. A typical state of the polarization resolved laser emissions in this case is shown in Figure 5.15. The results implies that the formation of the coupled dark-bright solitons is independent on the sign of the net cavity GVD.

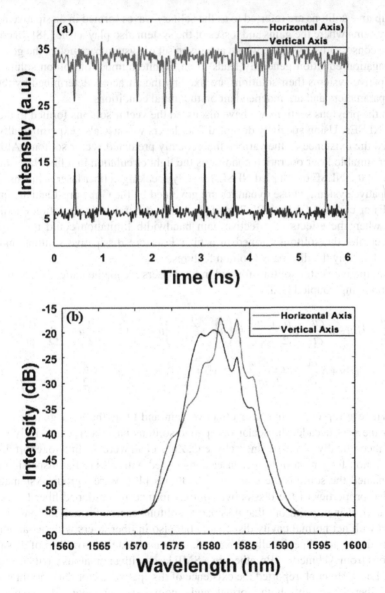

**FIGURE 5.15** The same as in Figure 5.14, but for the net normal-cavity-GVD regime. (Reprinted with permission from Hu et al. (2019) [37] © The Optical Society.)

We further checked the relative intensity of the dark and bright components in each complex by measuring the total laser output. It was observed that in the net anomalous GVD regime, the total laser emission is a bright pulse train determined by the bright solitons, while in the case of normal GVD it is a dark pulse train controlled by the dark solitons. The experimental results agree well with the theoretical predictions.

## 5.4 OUTLOOK ON DISSIPATIVE VECTOR SOLITONS

Dissipative solitons are referred to as the solitary waves formed in dissipative nonlinear systems where the gains and losses of the system also play a role [38]. In contrast to the conservative systems, where at a given set of system parameters the governing propagation equation generally leads to an infinite number of soliton solutions, in dissipative systems their solutions are fixed by the attractors determined by the system parameters and are independent on the initial conditions.

In the previous sections we have discussed the vector solitons formed in the coupled NLSEs. Using specially designed fiber lasers we have also experimentally confirmed the existence of the various theoretically predicted vector solitons. Although under suitable laser operation conditions the light circulation in a fiber laser mimics those of the NLSE or coupled NLSEs, strictly speaking, a fiber laser is intrinsically a dissipative system, whose dynamics are governed by the Ginzburg-Landau equation (GLE) or coupled GLEs [39]. Therefore, under general laser operation conditions, e.g., when the effects of effective gain bandwidth limitation could no longer be neglectable, the solitons formed in a fiber laser are dissipative solitons and they should display the dissipative soliton features.

Dissipative vector solitons formed in fiber lasers are mathematically described by the following coupled GLEs,

$$\frac{\partial u}{\partial z} = i\beta u - \delta \frac{\partial u}{\partial t} - i\frac{\beta_2}{2}\frac{\partial^2 u}{\partial t^2} + i\gamma\left(|u|^2 + \frac{2}{3}|v|^2\right)u + i\frac{\gamma}{3}v^2 u^* + \frac{g}{2}u + \frac{g}{2\Omega_g^2}\frac{\partial^2 u}{\partial t^2}$$

$$\frac{\partial v}{\partial z} = -i\beta v + \delta \frac{\partial v}{\partial t} - i\frac{\beta_2}{2}\frac{\partial^2 v}{\partial t^2} + i\gamma\left(|v|^2 + \frac{2}{3}|u|^2\right)v + i\frac{\gamma}{3}u^2 v^* + \frac{g}{2}v + \frac{g}{2\Omega_g^2}\frac{\partial^2 v}{\partial t^2} \quad (5.25)$$

Where $g$ represents the laser effective gain and $\Omega_g$ is the laser gain bandwidth. Over the past decade, the scalar dissipative solitons have been extensively studied both theoretically and experimentally, e.g., Zhao et al. were the first to report dissipative soliton formation in a fiber laser constructed with all normal dispersion fibers and named the solitons the gain guided solitons [40], Wise's group systematically studied properties of the dissipative solitons formed in Yb-doped fiber lasers [41]. However, it is to point out that dissipative solitons can be formed not only in fiber lasers with net normal cavity dispersion, but also in fiber lasers with net anomalous cavity dispersion. Specifically, Tang et al. demonstrated the formation of dissipative solitons from Akhmediev Breathers in a SMF laser with anomalous cavity dispersion [42]. Later, Hu et al. reported the existence of dissipative vector dark-bright solitons in a fiber laser with both normal and anomalous net cavity dispersion [43].

It has been shown that dissipative solitons could exhibit remarkable distinct properties from those of the NLSE type of solitons. Given that CGLEs model the dynamics of a variety of real-world nonlinear physical systems, we believe the studies on dissipative solitons are not only of fundamental importance but also hold promise potential for applications.

## 5.5 CONCLUSIONS

In the chapter we have investigated, both theoretically and experimentally, the various types of vector solitons formed in SMFs. Due to the vector nature of light, optical pulse propagation in a linear birefringent SMF is governed by the coupled NLSEs that admit different types of vector soliton solutions. Specifically, if the fiber birefringence is weak, the XPC between the two orthogonal polarization components of the light pulse is coherent. Under coherent XPC, a group of phase-locked vector solitons, that is, the polarization-maintained vector bright solitons, vector black solitons, and black-white vector solitons, could be formed. If the linear birefringence of the SMFs is strong, the XPC is incoherent. Under incoherent XPC, the group velocity locked bright-bright, dark-dark, and dark-bright solitons could be formed. Based on the dynamic analogues between solitons formed in specially designed fiber lasers and those of solitons formed through nonlinear pulse propagation in SMFs, we experimentally confirmed the existence of all the types of vector solitons theoretically predicted. Among the different types of vector solitons, the formation of both the coherently and incoherently coupled dark-bright solitons is of particular interest. The solitons could be formed both in the normal and anomalous dispersion regimes. As no scalar dark (bright) solitons could be formed in the anomalous (normal) dispersion SMFs, physically, they constitute a new form of optical solitons purely for the coupled nonlinear systems. In contrast to the dark-bright vector solitons, the other types of the formed vector solitons could also be considered as a bound state of the two orthogonally polarized scalar solitons. Their formation is a result of the soliton interaction. Since a fiber laser is intrinsically a dissipative system, its general dynamics is governed by the GLE or coupled GLEs, we anticipated that in addition to the various NLSE types of vector solitons, vector dissipative solitons could also be revealed in the system. Therefore, a fiber laser is an ideal testbed for the experimental study of the various forms of solitons of the NLSE, GLE or their coupled nonlinear systems.

## REFERENCES

[1] J. L. Hammack and H. Segur, "The Korteweg-de Vries equation and water waves. Part 2: Comparison with experiments," *J. Fluid Mech. 65*, 289–314 (1974).

[2] A. Chabchoub, O. Limmoun, H. Branger, N. Hoffmann, D. Proment, M. Onorato, and N. Akhmediev, "Experimental observation of dark solitons on the surface of water," *Phys. Rev. Lett. 110*, 124101 (2013).

[3] E. A. Kuznetsov, A. M. Rubenchik, V. E. Zakharov, "Soliton stability in plasma and hydrodynamics," *Phys. Rep. 142*, 103 (1986).

[4] K. E. Strecker, G. B. Partridge, A. G. Truscott, and R. G. Hulet, "Bright matter wave solitons in Bose–Einstein condensates," *New J. Phys 5*, 73 (2003).

[5] L. Brizhik, A. Eremko, B. Piette, W. Zakrzewski, "Solitons in alpha-helical proteins," *Phys. Rev. E. 70*, 031914 (2004).

[6] G. I. Stegemann and M. Segev, "Optical spatial solitons and their interactions: Universality and diversity," *Science 286*, 1518 (1999).

[7] Y. S. Kivshar and G. P. Agrawal, *Optical Solitons: From Fibers to Photonic Crystals* (Academic Press, San Diego, 2003).

[8] C. Sulem and P. L. Sulem, *The Nonlinear Schrödinger Equations* (Springer Verlag, New York, 1999).

[9] G. P. Agrawal, *Nonlinear Fiber Optics* (Academic Press, San Diego, 2012).

[10] L. F. Mollenauer, R. H. Stolen, and J. P. Gordon, "Experimental observation of picosecond pulse narrowing and solitons in optical fibers," *Phys. Rev. Lett. 45*, 1095 (1980).

[11] P. Emplit, J. P. Hamaide, F. Reynaud, C. Froehly, and A. Barthelemy, "Picosecond steps and dark pulses through nonlinear single mode fibers," *Opt. Commun. 62*, 374 (1987).

[12] C. R. Menyuk, "Nonlinear pulse propagation in birefringent optical fibers," *IEEE J. Quant. Electron. 23*, 174–176 (1987).

[13] N. N. Akhmediev, A. V. Buryak, J. M. Soto-Crespo, and D. R. Andersen, "Phase-locked stationary soliton states in birefringent nonlinear optical fibers," *J. Opt. Soc. Am. B 12*, 434–439 (1995).

[14] M. Haelterman, and A. P. Sheppard, "Polarization domain walls in diffractive or dispersive Kerr media," *Opt. Lett. 19*, 96–98 (1994).

[15] D. N. Christodoulides, "Black and white vector solitons in weakly birefringent optical fibers," *Phys. Lett. A 132*, 451 (1988).

[16] M. Lisak, A. Hook, and D. Anderson, "Symbiotic solitary-wave pairs sustained by cross-phase modulation in optical fibers," *J. Opt. Soc. Am. B 7*, 810–814 (1990).

[17] V. V. Afanasjev, "Soliton polarization rotation in fiber lasers," *Opt. Lett. 20*, 270–272 (1995).

[18] Y. F. Song, H. Zhang, D. Y. Tang, and D. Y. Shen, "Polarization rotation vector solitons in a graphene mode-locked fiber laser," *Opt. Express. 24*, 27283–27289 (2012).

[19] C. R. Menyuk, "Stability of solitons in birefringent optical fibers. I: Equal propagation amplitudes," *Opt. Lett. 12*, 614–616 (1987).

[20] X. D. Cao and C. J. McKinstrie, "Solitary-wave stability in birefringent optical fibers," *J. Opt. Soc. Am. B 10*, 1202–1207 (1993).

[21] V. K. Mesentsev and S. K. Turitsyn, "Stability of vector solitons in optical fibers," *Opt. Lett. 17*, 1497–1499 (1992).

[22] Y. S. Kivshar, and S. K. Turitsyn, "Vector dark solitons," *Opt. Lett. 18*, 337–339 (1993).

[23] M. N. Islam, C. D. Poole, and J. P. Gordon, "Soliton trapping in birefringent optical fibers," *Opt. Lett. 14*, 1011–1013 (1989).

[24] D. Y. Tang, J. Guo, Y. F. Song, G. D. Shao, L. M. Zhao, and D. Y. Shen, "Temporal cavity soliton formation in an anomalous dispersion cavity fiber laser," *J. Opt. Soc. Am. B 31*, 3050–3056 (2014).

[25] D. Y. Tang, Y. F. Song, J. Guo, Y. J. Xiang, and D. Y. Shen, "Polarization domain formation and domain dynamics in a quasi-isotropic cavity fiber laser," *IEEE J. select. Topic Quant. Electron. 20*, 0901309 (2014).

[26] S. T. Cundiff, B. C. Collings, N. N. Akhmediev, J. M. Soto-Crespo, K. Bergman, and W. H. Knox, "Observation of polarization-locked vector solitons in an optical fiber," *Phys. Rev. Lett. 82*, 3988–3991 (1999).

[27] S. M. J. Kelly, "Characteristic sideband instability of periodically amplified average soliton," *Electron. Lett. 28*, 806 (1992).

[28] H. Zhang, D. Y. Tang, L. M. Zhao, and N. Xiang, "Coherent energy exchange between components of a vector soliton in fiber lasers," *Opt. Express. 17*, 12618–12623 (2008).

[29] X. Hu, J. Guo, L. M. Zhao, J. Ma, and D. Y. Tang, "Coherently coupled vector black solitons in a quasi-isotropic cavity fiber laser," *Opt. Lett. 23*, 6563–6566 (2020).

[30] J. Ma, G. D. Shao, Y. F. Song, L. M. Zhao, Y. J. Xiang, D. Y. Shen, M. Richardson, and D. Y. Tang, "Observation of dark-bright vector solitons in fiber lasers," *Opt. Lett. 9*, 2185–2188 (2019).

[31] L. M. Zhao, D. Y. Tang, H. Zhang, X. Wu, and N. Xiang, "Soliton trapping in fiber lasers," *Opt. Express. 16*, 9528–9533 (2008).

[32] A. M. Weiner, J. P. Heritage, R. J. Hawkins, R. N. Thurston, E. M. Kirschner, D. E. Leaird, and W. J. Tomlinson, "Experimental observation of the fundamental dark soliton in optical fibers," *Phys. Rev. Lett. 61*, 2445–2448 (1988).

[33] J. E. Rothenberg, and H. K. Heinrich, "Observation of the formation of dark soliton trains in optical fibers," *Opt. Lett. 17*, 261–263 (1992).

[34] D. Y. Tang, J. Guo, Y. F. Song, H. Zhang, L. M. Zhao, and D. Y. Shen, "Dark soliton fiber lasers," *Opt. Express. 22*, 19831–19837 (2014).

[35] G. D. Shao, J. Guo, X. Hu, Y. F. Song, L. M. Zhao, and D. Y. Tang, "Vector dark solitons in a single mode fibre laser," *Laser Phys. Lett. 16*, 085110 (2019).

[36] D. Y. Tang, J. Guo, Y. F. Song, L. Li, L. M. Zhao, and D. Y. Shen, "GHz pulse train generation in fiber lasers by cavity induced modulation instability," *Opt. Fiber Technol. 20*, 610–614 (2014).

[37] X. Hu, J. Guo, G. D. Shao, Y. F. Song, S. W. Yoo, B. A. Malomed, and D. Y. Tang, "Observation of incoherently coupled dark-bright vector solitons in single-mode fibers," *Opt. Express. 27*, 18311 (2019).

[38] P. Grelu, and N. Akhmediev, "Dissipative solitons for mode-locked lasers," *Nat. Photonics 6*, 84–92 (2012).

[39] J. M. Soto-Crespo, and L. Pesquera, "Analytical approximation of the soliton solutions of the quintic complex Ginzburg-Landau equation," *Phys. Rev. E 56*, 7288 (1997).

[40] X. Hu, J. Guo, G. D. Shao, Y. F. Song, L. M. Zhao, L. Li, and D. Y. Tang, "Dissipative dark-bright vector solitons in fiber lasers," *Phys. Rev. A 101*, 063807 (2020).

[41] W. H. Renninger, A. Chong, and F. Wise, "Dissipative solitons in normal-dispersion fiber lasers," *Phys. Rev. A 77*, 023814 (2008).

[42] D. Y. Tang, J. Guo, Y. F. Song, G. D. Shao, L. M. Zhao, and D. Y. Shen, "Temporal cavity soliton formation in an anomalous dispersion cavity fiber laser," *J. Opt. Soc. Am. B 31*, 3050–3056 (2014).

[43] X. Hu, J. Guo, G. D. Shao, Y. F. Song, L. M. Zhao, L. Li, and D. Y. Tang, "Dissipative dark-bright vector solitons in fiber lasers," *Phys. Rev. A 101*, 063807 (2020).

# 6 Vector Solitons in Figure-Eight Fiber Lasers

## Min Luo and Zhi-Chao Luo

## CONTENTS

6.1 Introduction .................................................................................................169
6.2 Figure-Eight Fiber Laser and Working Principle ........................................170
6.3 Vector Nature of Multi-Soliton Patterns ....................................................171
    6.3.1 Fundamental Vector Soliton .............................................................172
    6.3.2 Random Static Distribution of Vector Multiple Solitons .................172
    6.3.3 Vector Soliton Cluster and Soliton Flow .........................................175
6.4 Vector Dissipative Soliton Resonance ........................................................178
    6.4.1 Typical Pulse Spectrum Operating in DSR Region .........................178
    6.4.2 Vector Nature of the DSR Pulse .......................................................179
6.5 Noise-Like Pulse Trapping ..........................................................................181
    6.5.1 From Conventional Soliton to Noise-Like Pulse .............................181
    6.5.2 Noise-Like Pulse Trapping with a Wavelength Shift .......................184
6.6 Conclusion ...................................................................................................186
Acknowledgements ..............................................................................................186
References .............................................................................................................186

## 6.1 INTRODUCTION

Temporal solitons in optical fibers, first observed by Mollenauer et al. in 1980, have been the fascinating subject of numerous theoretical and experimental investigations over the past decades [1]. The nonlinear Schrödinger equation (NLSE) is commonly used to describe the dynamics of fiber soliton mathematically. The derivation of the equation assumes that the polarization direction of the optics pulse is simply on the fast axis or the slow axis, so the solution of the equation is usually likewise named as scalar soliton [2–4]. Unfortunately, this statement overlooks a crucial fact, where single-mode fiber (SMF) is employed to construct fiber laser that supports two orthogonal polarization modes. The solitons could exhibit complex polarization dynamics due to the small amounts of random birefringence in SMF. Accordingly, it would also be noteworthy to investigate the vector characteristics of optical soliton along the two polarization axes of SMF, namely, vector solitons [5, 6].

As we know, passively mode-locked fiber lasers have been regarded as excellent platforms to explore the generation and the dynamics of soliton pulses. Indeed, by designing a polarization-insensitive laser cavity and detecting two orthogonal polarization components of the soliton, various nonlinear dynamics have been observed in vector soliton fiber lasers, i.e., polarization-locked vector soliton (PLVS) [7–10],

DOI: 10.1201/9781003206767-6

polarization-rotation vector soliton (PRVS) [11–14], and group-velocity-locked vector solitons (GVLVSs) [15–18]. Hence, the polarization-insensitive real saturable
absorbers (SAs), such as semiconductor saturable absorber mirror (SESAM), nanotube, and graphene SA were primarily employed in fiber laser to investigate the vector solitons. Nevertheless, the operation regimes of mode-locked fiber lasers with
real SAs are difficult to be tuned since it is insensitive to the adjustments of polarization controllers (PCs) inside the laser cavity. Again, the damage threshold is relatively close to the continuous mode-locking threshold, which is easy to break the
stable pulse state. Consequently, the mode-locked fiber lasers based on real SAs are
typically not suitable for observing non-conventional types of optical solitons, which
requires high pump power and specific polarization settings.

The charm of soliton dynamics lies in its complexity and diversity. Therefore, it is
also of great significance for the study of non-conventional solitons, such as multi-
soliton patterns [19–21], noise-like pulse trapping [22–25], and dissipative soliton
resonance (DSR) [26–29], etc. Regarding the experimental demonstrations of the
above-mentioned soliton nonlinear phenomena, most of them were observed in nonlinear polarization rotation (NPR)-based fiber lasers. Since a polarizer is required in the
fiber ring laser based on the NPR technique, it is not suitable for generating vector
solitons. Therefore, a question naturally arises as to whether these non-conventional
soliton nonlinear phenomena could be obtained in fiber lasers based on other mode-
locking techniques. Satisfyingly, due to its structural characteristics, the figure-eight
fiber laser could be also utilized to investigate the vector characteristics of various
soliton types [21, 25, 29]. One advantage that deserves to be highlighted is that there
were no polarization-loss dependent components (i.e., polarization-sensitive isolator).
And, more importantly, the intra-cavity polarization controller (PC) offers one more
degree of freedom to adjust the operation regime in the figure-eight fiber laser based on
a nonlinear amplifier loop mirror (NALM) or a nonlinear optical loop mirror (NOLM).

This chapter reviews the progress of vector features of different soliton types in
figure-eight fiber lasers. First, we briefly introduced the structure of the figure-eight
fiber laser and its working principle. In the next section, we will describe experimentally the fundamental vector soliton and various vector multi-soliton patterns (i.e.,
random static distribution of vector multiple solitons, vector soliton cluster, and vector soliton flow). By altering one or several parameters (i.e., the cavity length, pump
power), we then achieved the mode-locked pulse with the pulse profile transition
from sech-like to rectangular in the DSR region. The rectangular pulse trapping of
two orthogonal polarization components centered at different wavelengths was
observed. Finally, we focused on the trapping of a noise-like pulse in a passively
mode-locked figure-eight fiber laser. Despite the noise-like operation regime, it was
found that the pulse trapping of two polarization components centered at different
wavelengths was still obtained. They propagated along the laser cavity as a group-
velocity locked vector soliton (GLVS).

## 6.2   FIGURE-EIGHT FIBER LASER AND WORKING PRINCIPLE

Figure 6.1 exhibits the schematic of the figure-eight fiber laser. It is based on a
nonlinear amplifying loop mirror (NALM) coupled to a unidirectional ring cavity

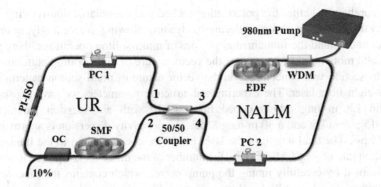

**FIGURE 6.1** Schematic of the polarization-insensitive figure-eight fiber laser. EDF, Erbium-doped fiber; WDM, Wavelength-division multiplexer; PI-ISO, Polarization-insensitive optical isolator; OC, Optical coupler; PC, Polarization controller.

through a 50/50 fiber coupler. Here, a piece of erbium-doped fiber (EDF) is used as the gain medium, pumped by a 980 nm laser diode. Two polarization controllers (PCs) are introduced to adjust the cavity birefringence. A polarization-independent isolator (ISO) ensures the unidirectional operation. A 10% fiber coupler is used to output the laser. In addition, a polarization beam splitter (PBS) is connected to the output coupler to resolve the two orthogonal polarization components.

The working principle of the figure-eight fiber laser can be expressed through the following process. The 50/50 fiber coupler connecting the left and right circles divides the incident light at port 1 into two beams of equivalent amplitude, which pass via the straight arm and the cross arm of the coupler, respectively. Any beam passing through the cross arm will produce a phase shift of $\pi/2$. Since the erbium-doped is near the coupler, the clockwise propagating light is amplified foremost, and the counterclockwise will not be strengthened until it escapes the cavity. Two beams of light propagating in different directions travel around the NALM cavity to obtain extra nonlinear phase shifts. As long as the two beams of light are realized with a phase shift difference of $\pi$, plus two phase shifts through the cross arms, the total phase shift is $2\pi$, and the light will interfere at port 2. In this case, the output change from port 1 to port 2 is realized; that is, the function of the optical switch is realized. In other words, if the phase shift of the central part with the high intensity of the pulse adjusted by NALM is close to $\pi$, the central part of the pulse will be transmitted. Still, the light intensity of the leading and trailing edges of the pulse is weak, leading to a small phase shift and reflection by self-phase modulation (SPM). The overall result is that the input pulse is narrowing during the propagation inside the NALM cavity.

## 6.3   VECTOR NATURE OF MULTI-SOLITON PATTERNS

The vector nature of multi-soliton patterns is worthwhile for indicating the intrin-sically physical features of multi-soliton dynamics. For instance, as a type of the multi-soliton pattern, soliton flow or soliton cluster in a fiber laser is a typical dynamic process. From the viewpoint of fundamental physics, it would be attractive to comprehend whether the polarization-locked and the polarization-rotating vector soliton flow could be observed during its dynamic flowing. Accordingly, in order to further understand the fundamental physics of multi-soliton operations, there would be a solid motivation to investigate the vector nature of multi-soliton patterns.

This section will concentrate on the vector nature of multi-soliton patterns in the figure-eight fiber laser. The experimental structure parameters we have chosen here contain 1.8 m-long erbium-doped fiber (EDF) with a dispersion parameter of $D = -15$ ps/nm/km and a 30 m-long SMF. The net cavity dispersion is approximately $-1.125$ ps$^2$. The total length of the laser is ~58 m, corresponding to the fundamental repetition rate of ~3.64 MHz. Here, the number of the mode-locked pulses was able to be governed by carefully tuning the pump power, which contains the random static distribution of vector multiple solitons, vector soliton cluster, and vector soliton flow. It should be noted that no polarizer was utilized in the cavity, and all fibers employed had weak birefringence. Therefore, vector solitons were thus naturally acquired. By using polarization-resolved measurement, both the fundamental polarization-locked vector soliton (PLVS) and polarization-rotating vector soliton (PRVS) can be identified [21].

### 6.3.1 FUNDAMENTAL VECTOR SOLITON

Firstly, we compare the vector nature of fundamental PLVS and PRVS in Figures 6.2 and 6.3. The transition between the two operation regimes can be accomplished by slightly adjusting the PC. Figure 6.2a depicts the temporal waveforms measured without and with passing through a PBS, respectively. All the solitons passing after the PBS simultaneously have the same pulse intensity [30]. However, for the single PRVS (Figure 6.3a), being different from the state of fundamental PLVS, the pulse intensities of the two polarization components varied periodically with the twice cavity roundtrip times. Figure 6.2b presents the corresponding spectra of PLVS. The spectra exhibit Kelly sidebands, confirming that the mode-locked pulse is the solitary wave of a fiber laser with an anomalous dispersion regime [31]. Then, Figure 6.3b depicts the corresponding optical spectra of the fundamental PRVS. Comparing with the optical spectra of the fundamental PLVS, there is an extra set of spectral sidebands on the polarization resolved spectra. The set of optical sidebands is similar to the ordinary Kelly sidebands before an external PBS. Distinctly, the intensities of the two optical sidebands alternated between two values in the two polarization components when measured after the PBS. Here, this kind of spectral sidebands is a special type of Kelly sidebands, which are formed due to a locally destructive interference between the resonance-enhanced dispersive waves and solitons [32]. In addition, it is also believed that the locations of the additionally spectral sidebands of PRVS are related to the rotation period of PRVS in the laser cavity [14, 30]. The Figures 6.2c and 6.3c show the autocorrelation trace of total pulse in fundamental PLVS and PRVS, respectively.

### 6.3.2 RANDOM STATIC DISTRIBUTION OF VECTOR MULTIPLE SOLITONS

The various multiple solitons patterns can be obtained by gradually increasing the pump power and adjusting the intra-cavity PCs. One of the most typical states is

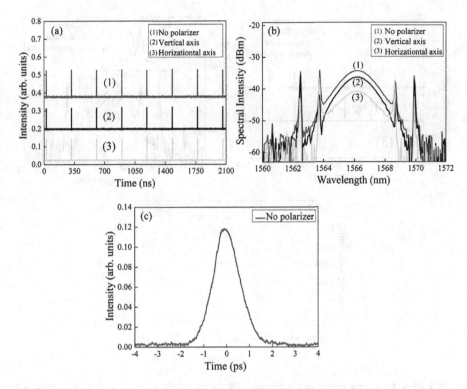

**FIGURE 6.2** Vector nature of fundamental PLVS. (a) Oscilloscope traces. (b) Polarization-resolved spectra. (c) Autocorrelation trace of total pulse. (Adapted from Ning et al. (2014) [21].)

that several stable vector solitons are randomly distributed over the cavity under a pump power of ~49 mW. The state of random static distribution of multiple PLVSs is shown in Figure 6.4. According to PLVSs characteristics, uniform polarization would remain consistent when propagating in the cavity. Indeed, all the solitons have the same pulse intensity after every roundtrip when passing through the PBS, as illustrated in Figure 6.4a. Figure 6.4b shows the corresponding optical spectra. We need to highlight a typical characteristic here: apart from the Kelly sidebands, another set of optical sidebands (indicated by black arrows) in the polarization-resolved spectra. In addition, these special sidebands present peak-dip alteration between the two polarization axes. Referring to [33], we assume the formation resulting from the four-wave-mixing coupling (also called coherent energy exchange) between the two polarization components of the PLVSs. Due to the different formation mechanisms of the Kelly sidebands, a fantastic phenomenon can be discovered. That is, the location of the additionally spectral sidebands of PLVS could be dynamically altered by adequately adjusting the PCs while the Kelly sidebands remained in the original positions.

The multiple PRVSs of the state of random static distribution could be further obtained by rotating the paddles of intra-cavity PCs, as shown in Figure 6.5. As can

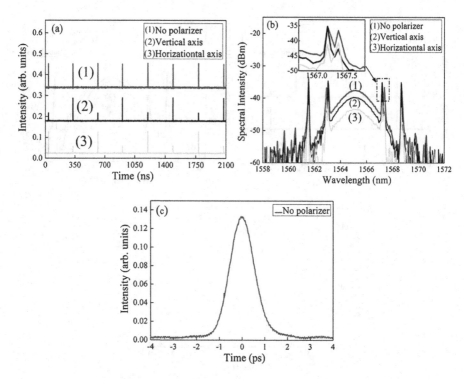

**FIGURE 6.3** Vector nature of fundamental PRVS. (a) Oscilloscope traces. (b) Polarization-resolved spectra. (c) Autocorrelation trace of total pulse. (Adapted from Ning et al. (2014) [21].)

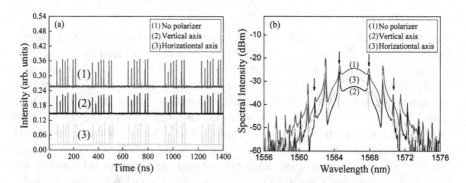

**FIGURE 6.4** The state of random static distribution of multiple PLVSs. (a) Oscilloscope traces. (b) Polarization-resolved spectra. (Adapted from Ning et al. (2014) [21].)

be seen in Figure 6.5a, the intensities of all the pulses in the oscilloscope trace are equal without passing through the PBS. But the intensities of pulses subsequently altered between two values after passing through the PBS, suggesting that the soliton polarization is rotating in the cavity. The soliton polarization rotation was locked to twice the cavity roundtrip time, indicating that intensity and polarization state of the

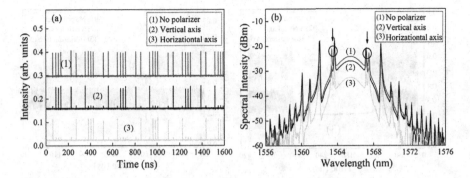

**FIGURE 6.5**   The state of random static distribution of multiple PRVSs. (a) Oscilloscope traces. (b) Polarization-resolved spectra. (Adapted from Ning et al. (2014) [21].)

vector soliton returned to its original value every two-cavity roundtrip time. In addition, other periods of soliton polarization rotation could be acquired by carefully adjusting the cavity parameters. Figure 6.5b shows the corresponding spectra of the PRVSs. Significantly different from the optical spectra of PLVS (see Figure 6.4b), two extra sets of spectral sidebands (indicated by black arrows) are caused by the periodic polarization rotation of the vector solitons.

### 6.3.3   Vector Soliton Cluster and Soliton Flow

When increasing the pump power to ~155 mW, the number of solitons increased, and the solitons tended to be dynamic. In this case, the solitons moved relative to each other and came concurrently. The solitons would form a soliton cluster in which the solitons bunched together tightly just by adjusting the PC settings, as shown in Figure 6.6 and 6.7. Figure 6.6 illustrates the polarization-locked vector soliton cluster (PLVSC). Figures 6.6a and 6.6b show the pulse-trains of PLVSC. All the envelope widths of the soliton clusters are uniform both in the total output and in the polarization resolved outputs (Figure 6.6a). Figure 6.6b illustrates oscilloscope traces of single soliton packet with small span in detail. The pulse trains are consistent among the total pulses and the two orthogonal polarization components. Figure 6.6c shows the autocorrelation trace of pulse before the PBS. Similar to Figure 6.4b, the Kelly sidebands and peak-dip optical sidebands on the spectra also coexist, as shown in Figure 6.6d. Most noteworthy is that there is no modulation on the mode-locked spectrum. It is because the large temporal separations among these solitons inside the clusters lead to no evident interference patters on the spectrum. Consequently, the PLVSC is somewhat different from the condensed soliton phase, consisting of tens or hundreds of phase-locked solitons and evident spectral modulations [34, 35].

The polarization rotation vector soliton cluster (PRVSC) was achieved by further adjusting the PCs. Figure 6.7a shows the pulse trains of PRVSC. Figure 6.7b further shows the intensities of solitons vary between the two orthogonal polarization components in detail. Moreover, we also measured the autocorrelation trace of the PRVSC, as presented in Figure 6.7c. Figure 6.7d shows the corresponding spectra of

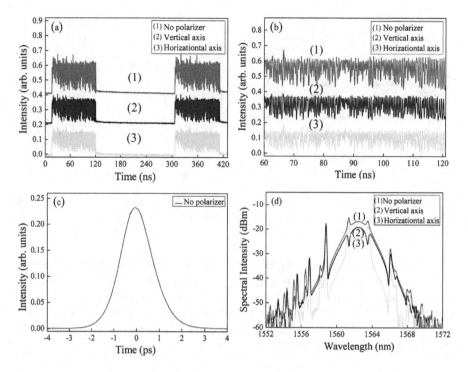

**FIGURE 6.6** Vector nature of the PLVSC. (a) Oscilloscope traces. (b) Corresponding oscilloscope traces with small span in detail. (c) Autocorrelation trace of total pulse. (d) Polarization resolved spectra. (Adapted from Ning et al. (2014) [21].)

PRVSC. Interestingly, more sets of the spectral sidebands (indicated by black arrows) appear on the spectra. This indicates that both the PRVSC and the PLVSC are distinct from the condensed soliton phase.

Next, when the pump power was further increased to ~204 mW, another intriguing dynamic multi-soliton pattern could be obtained: The vector soliton flow. This phenomenon can be expressed as the flowing solitons that arise from one soliton cluster and drift at different speeds to another soliton cluster. The soliton flow is a reverse phenomenon of the soliton rain, in which the pulses generated from the noisy and fall to the condensed phase or soliton cluster [36, 37]. Experimentally, increasing the pump power within a specific range made it easier to acquire more solitons from the soliton cluster. The higher speeds of these motion pulses could be observed. Figure 6.8 shows the case of the polarization-locked vector soliton flow (PLVSF). Figure 6.8a shows that all the pulses in the soliton cluster and the corresponding flow are uniform across the total output and the two polarization-resolved outputs. Clearly, this vector characteristic is consistent with conventional soliton. Figure 6.8b shows the spectral components of the PLVSF. Here, the observed extra spectral sidebands are caused by the four-wave-mixing effect, which is similar to PLVSs in Figure 6.4b.

The polarization-resolved vector soliton flow (PRVSF) was observed by adequately tuning the PCs, as revealed in the Figure 6.9. The intensities of all the flowing solitons at the two polarization axes varied twice the cycle of the fiber laser cavity

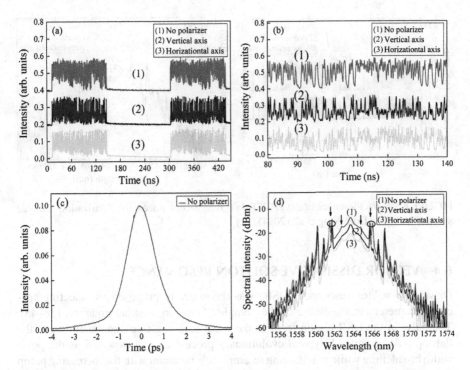

**FIGURE 6.7** Vector nature of the PRVSC. (a) Oscilloscope traces. (b) Corresponding oscilloscope traces with small span in detail. (c) Autocorrelation trace of total pulse. (d) Polarization resolved spectra. (Adapted from Ning et al. (2014) [21].)

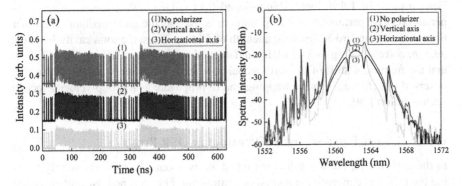

**FIGURE 6.8** Vector nature of the PLVSF. (a) Oscilloscope traces. (b) Polarization-resolved spectra. (Adapted from Ning et al. (2014) [21].)

(see Figure 6.9a). Yet, the polarization rotating characteristics of the soliton flow for the part of cluster could not be clearly distinguished. Figure 6.9b displays the spectra of PRVSF and the polarization-resolved components. Apparently, the extra sidebands caused by the periodic polarization rotation of vector solitons were also observed [30].

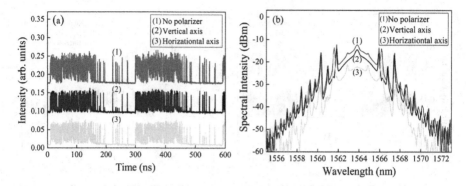

**FIGURE 6.9**   Vector nature of the PRVSF. (a) Oscilloscope traces. (b) Polarization-resolved spectra. (Adapted from Ning et al. (2014) [21].)

## 6.4   VECTOR DISSIPATIVE SOLITON RESONANCE

Dissipative soliton resonance (DSR) was theoretically proposed by selecting specific parameters in the frame of the complex Ginzburg-Landau equation [38–43]. The pulse in the DSR region features the wave breaking-free phenomenon and the flat-top pulse profile. A typical evolutionary process can be expected as the pulse width broadening while maintaining its amplitude invariant with the increasing pump power. In this case, the pulse energy can be significantly increased despite the overdriven intracavity nonlinear effect. This section will concentrate on the vector nature of rectangular pulse operating in dissipative soliton resonance (DSR) region in a nonlinear amplifier loop mirror (NALM)-based mode-locked figure-eight fiber laser. Here, a piece of 1.8 m long erbium-doped fiber (EDF) with a group velocity dispersion (GVD) parameter of −15 ps/nm/km is used as the gain medium. The other fibers are all standard SMFs with a length of 57.5 m. Thus, the total cavity length is 59.3 m, corresponding to 3.44 MHz fundamental repetition rate. Here, we will show that the frequency shift of two orthogonal polarized components of DSR pulse was observed to achieve the pulse trapping and to propagate as a group-velocity locked vector soliton [29].

### 6.4.1   TYPICAL PULSE SPECTRUM OPERATING IN DSR REGION

In the experiment, the mode-locking threshold is about 150 mW. By simply rotating the PCs, the conventional soliton operation could be obtained. Nevertheless, the multi-pulse operation would be observed in this case if the pump power was high enough. With further proper adjustment of the PCs, the pulse became breaking free and the pulse width broadened with the increasing pump power, which is the typical signature of the DSR phenomenon. The pulse spectrum operating in DSR region at the pump power of 250 mW (see Figure 6.10a). Figure 6.10b delivers the mode-locking pulse trains recorded by the oscilloscope at the pump power of 190 mW (solid line) and 250 mW (dotted line). Here, the pulse profiles are rectangular. Moreover, the pulse width increased from 41.7 ns to 58.1 ns when the pump power was adjusted

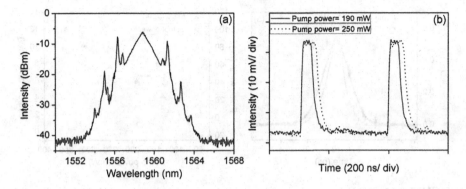

**FIGURE 6.10** (a) Typical spectrum of DSR pulse. (b) Pulse trains under the pump power of 190 mW and 250 mW. (Adapted from Luo et al. (2013) [29].)

from 190 mW to 250 mW. These observations demonstrate that the pulse obtained in our fiber laser operates in the DSR region.

As mentioned above, the self-starting mode-locking threshold is 150 mW. Here, one phenomenon we need to point out is that the fiber laser could sustain a mode-locking state when the pump power was decreased to ~70 mW due to the pump hysteresis [44]. We could notice the pulse profile evolution with the decreasing pump power, as illustrated in Figure 6.11a. The pulse profile transition from rectangular to sech-like shape was observed with the pump power from 90 mW to 70 mW. Likewise, the measured pulse duration was decreased from 5.1 ps to 0.97 ps. Further, we have displayed the measured pulse width and output power versus the pump power in Figure 6.11b. The output power of a rectangular pulse is 9.01 mW. In addition, the pulse width could be varied from 0.97 ps to 100.63 ns by enhancing the pump power from 70 to 350 mW.

## 6.4.2 Vector Nature of the DSR Pulse

The vector soliton generation is an intrinsic feature in our figure-eight fiber laser due to polarization-insensitive. Firstly, we set a proper pump power of 100 mW. Figure 6.12a illustrates the spectrum of vector soliton and the corresponding two orthogonal polarization components under the 100 mW pump power. The two orthogonal polarization components are discovered at distinct central wavelengths with a separation of 0.26 nm, whose intensity difference is ~6.1 dB. Hence, although the fiber birefringence exists in the laser cavity, the two polarization components could compensate for the fiber birefringence-induced polarization dispersion by shifting the center frequencies. Then they could trap each other as a group-velocity locked vector soliton. Here, the rectangular pulse trapping in the DSR region is similar to that of the conventional vector soliton, confirming that pulse trapping is a universal phenomenon of vector soliton despite the different formation mechanisms. Figure 6.12b depicts the vector nature of DSR in the time domain. The pulse profiles of two polarization components were both sech-like by employing a commercial autocorrelator. The horizontal and vertical pulse widths are both about 9.3 ps.

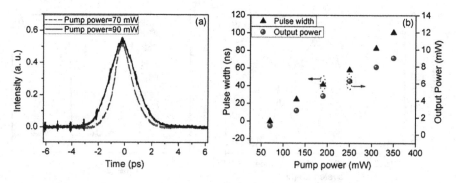

**FIGURE 6.11**    (a) Pulse profile evolution with decreasing pump power. (b) Pulse width and output power versus pump power. (Adapted from Luo et al. (2013) [29].)

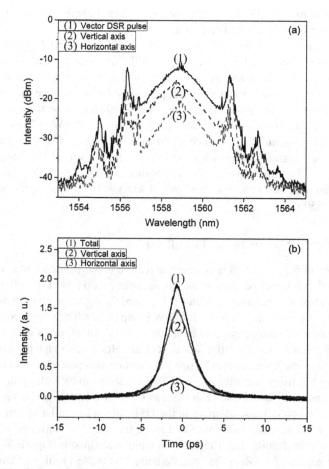

**FIGURE 6.12**    Vector nature of DSR pulse at the pump power of 100 mW. (a) Polarization-resolved spectra. (b) Autocorrelation traces. (Adapted from Luo et al. (2013) [29].)

Next, we will compare the vector nature of the pulse in the DSR region at different pump power levels. Typically, Figure 6.13 illustrates the vector nature of DSR pulse at the pump power of 300 mW. Differently, the spectral intensity difference of the two polarization components is larger than that at 100 mW pump power, which is ~17.2 dB, as shown in Figure 6.13a. In addition, the center wavelength shift is not so noticeable, being only ~0.1 nm. Since the polarization state of the vector DSR pulse evolved along with the fiber, the spectral intensities of two polarization components at the output port are related to the polarization state at the input port of PBS. The distinction in vector spectra between the cases of 100 mW and 300 mW is primarily due to the change of polarization state at the input port of PBS induced by the increased pump power. Further, the vector feature in the time domain was investigated. As expected, pulse profiles of the two polarization components are rectangular and maintain almost the same duration of ~75 ns (full width at half maximum), as displayed in Figure 6.13b. The above experimental results proved that two orthogonal polarization components could operate in the DSR region with the increasing pump power. The discussion here mainly concentrates on group-velocity locked vector soliton operating in the DSR region. Additional vector soliton dynamics of DSR such as PLVS, PRVS and coherent energy exchange may be acquired by further adjusting the cavity parameters.

## 6.5 NOISE-LIKE PULSE TRAPPING

The noise-like pulse is a localized wave packet that consists of multiple chaotic pulses with high peak powers [22–25]. Moreover, the noise-like pulse possesses a broadband and smooth averaged mode-locked spectrum. Accordingly, considering the unique characteristics of noise-like pulse in both time and spectral domains, it would be interesting to investigate the vector nature of a noise-like pulse. In this section, the figure-eight fiber laser includes a piece of 4 m long erbium-doped fiber (EDF), which is used as the gain medium. The other fibers are all standard SMFs with a length of 32.8 m. Thus, the fundamental repetition rate is 5.58 MHz [25].

### 6.5.1 FROM CONVENTIONAL SOLITON TO NOISE-LIKE PULSE

In the experiment, we firstly investigated the pulse trapping of conventional soliton operation under the pump power of ~100 mW. The measured results are summarized in Figure 6.14. The mode-locked spectrum with and without passing the PBS is exhibited (see Figure 6.14a). As can be seen here, the center wavelength of the mode-locked spectrum is 1575.4 nm. And the 3-dB spectral bandwidth is 3.2 nm. When the PBS resolved the mode-locked spectrum, it can be noticed that the two orthogonal polarization components are located at different wavelengths with a separation of 0.54 nm. The fiber birefringence-induced polarization dispersion could be compensated by shifting the center wavelengths. Then the pulse trapping was achieved and the two orthogonal polarization components propagated as a stable GLVS. The corresponding pulse trains are shown in Figure 6.14b. Figure 6.14c illustrates the autocorrelation trace of the mode-locked pulse. The pulse duration is measured to be ~0.9 ps. Thus, the time-bandwidth product is 0.348, indicating the pulse is slightly

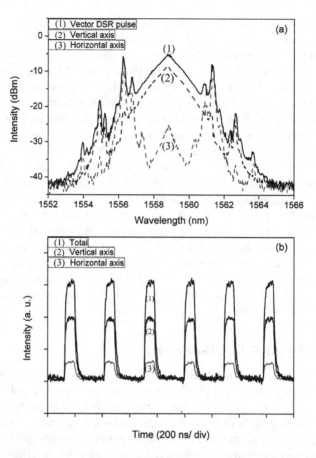

**FIGURE 6.13**  Vector characteristics of DSR pulse at the pump power of 300 mW. (a) Polarization resolved spectra. (b) Oscilloscope traces. (Adapted from Luo et al. (2013) [29].)

chirped. It should be mentioned that the pulse durations of two polarization components are somewhat dissimilar from each other, which is yielded by the different bandwidths of mode-locked spectra.

To achieve the transformation from conventional soliton to noise-like pulse, we carefully rotate the orientations of PCs. In this case, the mode-locked spectrum broadened suddenly, and the Kelly sidebands disappeared simultaneously. From the spectral characteristics, it can be deduced that the obtained pulse is operating in a noise-like regime. The mode-locked spectrum of noise-like pulse is shown in Figure 6.15a. The corresponding pulse-train is measured in Figure 6.15b. Here, the pulse is similar to those of conventional solitons due to the limited bandwidth of oscilloscope. The autocorrelation trace of the mode-locked pulse was measured to identify further that the fiber laser operated in noise-like pulse regime. The autocorrelation traces are presented in Figure 6.15c, in which we can notice that there is a narrow coherent peak with broad shoulders. The pulse characteristics in spectral and time domains agree with the intrinsic characteristics of the previously reported noise-like pulse fiber laser [22–25].

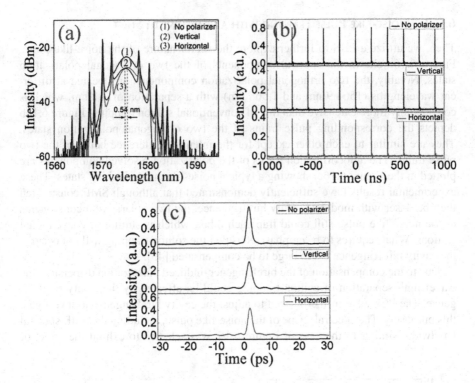

**FIGURE 6.14** Conventional soliton operation. (a) Polarization-resolved spectra. (b) Corresponding pulse trains. (c) Autocorrelation traces. (Adapted from Luo et al. (2015) [25].)

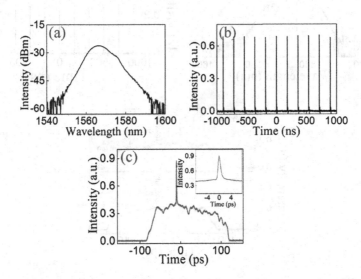

**FIGURE 6.15** Noise-like pulse operation. (a) Spectrum. (b) Pulse train. (c) Autocorrelation traces. (Adapted from Luo et al. (2015) [25].)

## 6.5.2 NOISE-LIKE PULSE TRAPPING WITH A WAVELENGTH SHIFT

Then, we utilize a PBS to further explore the vector nature of the noise-like pulse. Figure 6.16a shows the spectral components of the two orthogonal polarization states. Notably, the two orthogonal polarization components are located at different wavelengths (1564.9 nm and 1,568 nm) with a separation of 3.1 nm, which is considerably more extensive than that of conventional soliton trapping. Figure 6.16b depicts the corresponding pulse trains of the two orthogonal polarization states. They are similar to each other except for the intensity difference between the two pulse trains. The autocorrelation traces of the polarization-resolved components are plotted in the Figure 6.16c, showing a typical noise-like mode-locking states. These experimental results have sufficiently demonstrated that although SMF constructed the fiber laser with moderate cavity birefringence, the two polarization components of the noise-like pulse still could trap each other, which is similar to conventional solitons. What requires to be emphasized is that the soliton trapping will not occur if the cavity birefringence is too large to be compensated [45].

Due to the compensation of the birefringence-induced polarization dispersion, the wavelength separation of soliton trapping could be affected by the cavity birefringence. The PCs were further rotated to adjust the cavity birefringence to investigate this purposely. The spectral shape of the noise-like pulse, including the 3-dB spectral bandwidth similar to the case of Figure 6.15, was selected to exhibit the effect of

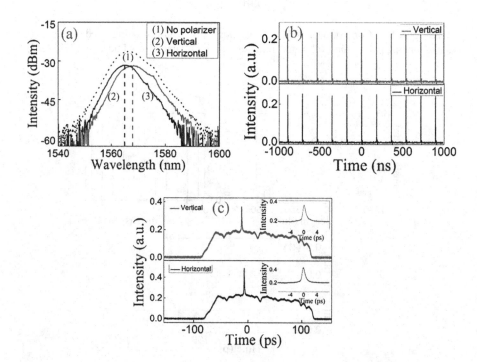

**FIGURE 6.16** Noise-like pulse trapping. (a) Polarization-resolved spectra. (b) Corresponding pulse trains. (c) Corresponding autocorrelation traces. (Adapted from Luo et al. (2015) [25].)

cavity birefringence on the wavelength shift nicely. Figure 6.17a offers the mode-locked spectrum with the center wavelength varying slightly to be 1567.5 nm. The corresponding pulse train is shown in Figure 6.17b. Again, the resolved optical spectra of the two components are shown in Figure 6.17c. The center wavelengths of the two polarization components are 1567.1 nm (blue curve) and 1571.9 nm (red curve), respectively. Thus, the wavelength shift of pulse trapping reached 4.8 nm. In this case, the intensity difference of the two polarization components is about 8.3 dB, which is larger than in Figure 6.16a. Note that that the considerable intensity difference between the two polarization components could be induced by the large ellipticity of vector noise-like pulse. Specifically, the nonlinear birefringence induced by the noise-like pulse could be much larger than that caused by the conventional solitons. Therefore, in the case of a noise-like operation regime in the fiber laser, the frequency shift of the two polarized components needs to be large enough to compensate for the fiber birefringence-induced polarization dispersion. Then, they could trap each other as a group-velocity locked vector soliton. Because of the high peak power of the noise-like wave packet, it is expected that the wavelength shift would be sensitive to the cavity birefringence, which was also verified by our experimental results. Figure 6.17d shows the resolved pulse trains centered at two wavelengths. Both of them keep uniform intensities, implying that the two polarization components yet could trap each other despite the variation of cavity birefringence.

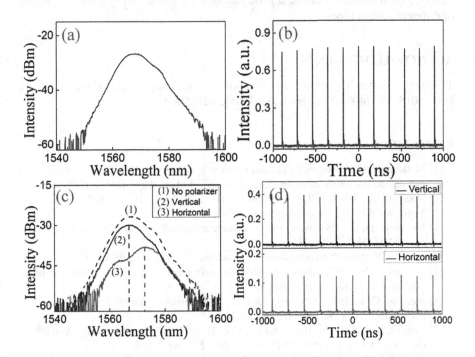

**FIGURE 6.17** Noise-like pulse trapping with a wavelength shift of 4.8 nm. (a) Spectrum of total pulse. (b) Total pulse train. Inset: Corresponding autocorrelation trace. (c) Polarization-resolved spectra. (d) Corresponding pulse trains. (Adapted from Luo et al. (2015) [25].)

## 6.6 CONCLUSION

This chapter has reviewed the recent progress of non-conventional vector solitons such as multi-soliton patterns, dissipative soliton resonance, and noise-like pulse trapping in figure-eight fiber lasers. We firstly presented the vector dynamics of multi-soliton patterns through tuning the pump power and adjusting the PCs. Here, it was shown that there were peak-dip optical sidebands on the spectra of every type of multiple PLVSs, indicating that the coherent energy exchange always existed between the two orthogonal polarization components of multiple PLVSs. As for the PRVSs, there were always extra sidebands on the spectra of the fundamental PRVS and every type of multiple PRVSs, whose formation was caused by the periodic polarization rotation of the vector solitons. We then introduced the vector characteristics of the pulse operating in the DSR region in a polarization-insensitive figure-eight fiber laser. The frequency shift of two orthogonal polarized components of DSR pulse was observed to achieve the pulse trapping and propagating as a group-velocity locked vector soliton. The spectral distributions of the two polarization components were different from each other. Finally, we have investigated the trapping of the noise-like pulse in a figure-eight fiber laser. The wavelength shift of noise-like pulse trapping was much larger than that of conventional soliton trapping, which could be up to 4.8 nm by properly rotating the PCs. In summary, the obtained results reveal the vector characteristics of non-conventional solitons in fiber lasers and further demonstrate that the figure-eight fiber laser could be a good platform for investigating the vector nature of different soliton types.

## ACKNOWLEDGEMENTS

This work was partially supported by National Natural Science Foundation of China (11874018, 11974006, 61805084, 61875058, 62175069).

## REFERENCES

[1] Mollenauer, L. F., Stolen, R. H., and Gordon, J. P. 1980. Experimental observation of picosecond pulse narrowing and solitons in optical fibers. *Phys. Rev. Lett.* 45 1095.

[2] Dodd, R. K., Eilbeck, J. C., Gibbon, J. D., and Morris, H. C. 1982. *Solitons and Nonlinear Wave Equations.* Academic Press.

[3] Peregrine, D. H. 1983. Water waves, nonlinear Schrödinger equations and their solutions *ANZIAM. J.* 25 16–43.

[4] Kodama, Y. 1985. Optical solitons in a monomode fiber. *J. Stat. Phys.* 39 597–614.

[5] Menyuk, C. R. 1987. Stability of solitons in birefringent optical fibers. I: Equal propagation amplitudes. *Opt. Lett.* 12 614–616.

[6] Menyuk, C. R. 1988. Stability of solitons in birefringent optical fibers. II. Arbitrary amplitudes. *J. Opt. Soc. Am. B* 5 392–402.

[7] Cundiff, S. T., Collings, B. C., and Knox, W. H. 1997. Polarization locking in an isotropic, modelocked soliton Er/Yb fiber laser. *Opt. Express* 1 12–20.

[8] Cundiff, S. T., Collings, B. C., Akhmediev, N. N., Soto-Crespo, J. M., Bergman, K., and Knox, W. H. 1999. Observation of polarization-locked vector solitons in an optical fiber. *Phys. Rev. Lett.* 82 3988.

[9] Tang, D. Y., Zhang, H., Zhao, L. M., and Wu, X. 2008. Observation of high-order polarization-locked vector solitons in a fiber laser. *Phys. Rev. Lett. 101* 153904.

[10] Mou, C., Sergeyev, S., Rozhin, A., and Turistyn, S. 2011. All-fiber polarization locked vector soliton laser using carbon nanotubes. *Opt. Lett. 36* 3831–3833.

[11] Afanasjev, V. V. 1995. Soliton polarization rotation in fiber lasers. *Opt. Lett. 20* 270–272.

[12] Zhao, L. M., Tang, D. Y., Wu, X. H., Zhang, H., and Tam, H. Y. 2009. Coexistence of polarization-locked and polarization-rotating vector solitons in a fiber laser with SESAM, *Opt. Lett. 34* 3059–3061.

[13] Zhang, H., Tang, D. Y., Wu, X., and Zhao, L. M. 2009. Multi-wavelength dissipative soliton operation of an erbium-doped fiber laser. *Opt. Express 17* 12692–12697.

[14] Song, Y. F., Zhang, H., and Tang, D. Y. 2012. Polarization rotation vector solitons in a graphene mode-locked fiber laser. *Opt. Express 20* 27283–27289.

[15] Islam, M. N., Poole, C. D., and Gordon, J. P. 1989. Soliton trapping in birefringent optical fibers. *Opt. Lett. 14* 1011–1013.

[16] Zhao, L. M., Tang, D. Y., Zhang, H., Wu, X. H., and Xiang, N. 2008. Soliton trapping in fiber lasers. *Opt. Express 16* 9528–9533.

[17] Mao, D., Liu, X. M., and Lu, H. 2012. Observation of pulse trapping in a near-zero dispersion regime. *Opt. Lett. 37* 2619–2621.

[18] Yuan, X. Z., Yang, T., Chen, J. L., He, X., Huang, H. C., Xu, S. H., and Yang, Z. M. 2013. Experimental observation of vector solitons in a highly birefringent cavity of ytterbium-doped fiber laser. *Opt. Express 21* 23866–23872.

[19] Tang, D. Y., Zhao, B., Zhao, L. M., and Tam, H. Y. 2005. Soliton interaction in a fiber ring laser. *Phys. Rev. E 72* 016616.

[20] Haboucha, A., Leblond, H., Salhi, M., Komarov, A., and Sanchez, F. 2008. Analysis of soliton pattern formation in passively mode-locked fiber lasers. *Phys. Rev. A 78* 043806.

[21] Ning, Q. Y., Liu, H., Zheng, X. W., Yu, W., Luo, A. P., Huang, X. G., Luo, Z. C., Xu, W. C., Xu, S. H., and Yang, Z. M. 2014. Vector nature of multi-soliton patterns in a passively mode-locked figure-eight fiber laser. *Opt. Express 22* 11900–11911.

[22] Horowitz, M., Barad, Y., and Silberberg, Y. 1997. Noiselike pulses with a broadband spectrum generated from an erbium-doped fiber laser. *Opt. Lett. 22* 799–801.

[23] Jeong, Y., Vazquez-Zuniga, L. A., Lee, S., and Kwon, Y. 2014. On the formation of noise-like pulses in fiber ring cavity configurations. *Opt. Fiber Technol. 20* 575–592.

[24] Kobtsev, S., Kukarin, S., Smirnov, S., Turitsyn, S., and Latkin, A. 2009. Generation of double-scale femto/pico-second optical lumps in mode-locked fiber lasers. *Opt. Express 17* 20707–20713.

[25] Luo, A. P., Luo, Z. C., Liu, H., Zheng, X. W., Ning, Q. Y., Zhao, N., Chen, W. C., and Xu, W. C. 2015. Noise-like pulse trapping in a figure-eight fiber laser. *Opt. Express 23* 10421–10427.

[26] Wu, X., Tang, D. Y., Zhang, H., and Zhao, L. M. 2009. Dissipative soliton resonance in an all-normal-dispersion erbium-doped fiber laser. *Opt. Express 17* 5580–5584.

[27] Duan, L., Liu, X., Mao, D., Wang, L., and Wang, G. 2012. Experimental observation of dissipative soliton resonance in an anomalous-dispersion fiber laser. *Opt. Express 20* 265–270.

[28] Luo, Z. C., Cao, W. J., Lin, Z. B., Cai, Z. R., Luo, A. P., and Xu, W. C. 2012. Pulse dynamics of dissipative soliton resonance with large duration-tuning range in a fiber ring laser. *Opt. Lett. 37* 4777–4779.

[29] Luo, Z. C., Ning, Q. Y., Mo, H. L., Cui, H., Liu, J., Wu, L. J., Luo, A. P., and Xu, W. C. 2013. Vector dissipative soliton resonance in a fiber laser. *Opt. Express 21* 10199–10204.

[30] Song, Y. F., Li, L., and Tang, D. Y. 2013. Quasi-periodicity of vector solitons in a graphene mode-locked fiber. *Laser Phys. Lett. 10* 125103.

[31] Kelly, S. M. J. 1992. Characteristic sideband instability of periodically amplified average soliton. *Electron. Lett. 28* 806–807.

[32] Du, Y. Q., Shu, X. W., Cao, H. R., and Cheng, P. Y. 2017. Dynamics of dispersive wave and regimes of different kinds of sideband generation in mode-locked soliton fiber lasers. *IEEE. J. Sel. Top. Quantum Electron. 24* 1–8.

[33] Zhang, H., Tang, D. Y., Zhao, L. M., and Xiang, N. 2008. Coherent energy exchange between components of a vector soliton in fiber lasers. *Opt. Express 16* 12618–12623.

[34] Amrani, F., Salhi, M., Grelu, P., Leblond, H., and Sanchez, F. 2011. Universal soliton pattern formations in passively mode-locked fiber lasers. *Opt. Lett. 36* 1545–1547.

[35] Amrani, F., Salhi, M., Leblond, H., Haboucha, A., and Sanchez, F. 2011. Intricate solitons state in passively mode-locked fiber lasers. *Opt. Express 19* 13134–13139.

[36] Chouli, S., and Grelu, P. 2009. Rains of solitons in a fiber laser. *Opt. Express 17* 11776–11781.

[37] Chouli, S., and Grelu, P. 2010. Soliton rains in a fiber laser: An experimental study. *Phys. Rev. A 81* 063829.

[38] Zhao, L. M., Tang, D. Y., Zhang, H., and Wu, X. 2008. Polarization rotation locking of vector solitons in a fiber ring laser. *Opt. Express 16* 10053–10058.

[39] Chang, W., Ankiewicz, A., Soto-Crespo, J. M., and Akhmediev, N. 2008. Dissipative soliton resonances. *Phys. Rev. A 78* 023830.

[40] Chang, W., Soto-Crespo, J. M., Ankiewicz, A., and Akhmediev, N. 2009. Dissipative soliton resonances in the anomalous dispersion regime. *Phys. Rev. A 79* 033840.

[41] Grelu, P., Chang, W., Ankiewicz, A., Soto-Crespo, J. M., and Akhmediev, N. 2010. Dissipative soliton resonance as a guideline for high-energy pulse laser oscillators. *J. Opt. Soc. Am. B 27* 2336–2341.

[42] Ding, E., Grelu, P., and Kutz, J. N. 2011. Dissipative soliton resonance in a passively mode-locked fiber laser. *Opt. Lett. 36* 1146–1148.

[43] Grelu, P., and Akhmediev, N. 2012. Dissipative solitons for mode-locked lasers. *Nat. Photonics 6* 84–92.

[44] Komarov, A., Leblond, H., and Sanchez, F. 2005. Multistability and hysteresis phenomena in passively mode-locked fiber lasers. *Phys. Rev. A 71* 053809.

[45] Wang, Y., Wang, S., Luo, J., Ge, Y., Li, L., Tang, D. Y., Shen, D. Y., Zhang, S. M., Wise, F. K., and Zhao, L. M. 2014. Vector soliton generation in a Tm fiber laser. *IEEE. Photon. Technol. Lett. 26* 769–772.

# 7 Polarization Dynamics of Mode-Locked Fiber Lasers with Dispersion Management

*Tao Zhu and Lei Gao*

## CONTENTS

7.1 Polarization Dynamics of Abnormal Dispersion Mode-Locked Lasers ...... 189
7.2 Polarization Dynamics of Normal Dispersion Mode-Locked Lasers .......... 190
7.3 Polarization Dynamics of Near-Zero Net Dispersion
Mode-Locked Lasers ............................................................................... 197
7.4 Polarization Dynamics of Partially Mode-Locked Fiber Lasers ................. 199
References ........................................................................................................ 202

## 7.1 POLARIZATION DYNAMICS OF ABNORMAL DISPERSION MODE-LOCKED LASERS

In general, the polarization-dependent mode-locking mechanism results in a significant restriction on the operation of vector solitons. However, recent advances have shown that vector solitons can still operate stably in polarization-dependent mode-locked lasers by inducing rapid polarization evolution within the laser cavity. Xiang et al. demonstrated a bidirectional fiber laser based on nonlinear polarization rotation (NPR) near a scalar vector operating state [1]. The test system is shown in Figure 7.1. In the clockwise (CW) direction of laser operation, NPR-based polarization-dependent effects dominate the evolution of linearly polarized pulses and limit the formation of vector solitons. This is reflected in the fact that the component of vertical projection can be almost completely suppressed by polarization beam splitter (PBS) through a properly adjusting polarization controller (PC3). In the counterclockwise (CCW) running direction, the strong local birefringence induced by the polarization-maintaining fiber promotes rapid polarization evolution, resulting in the formation of vector solitons. That is, the pulses cannot be completely suppressed and are easily detected. The modulation spectrum along the vertical axis is due to the interference introduced by the PBS-induced projection of two orthogonal polarization components along the horizontal and vertical axes.

DOI: 10.1201/9781003206767-7

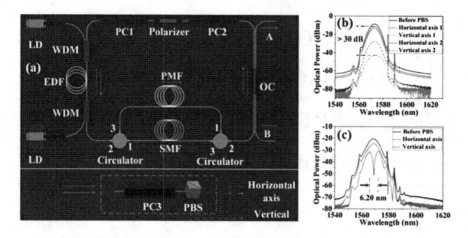

**FIGURE 7.1** (a) Schematic diagram of a bidirectional fiber laser. (b) Clockwise-running soliton output spectrum. (c) Counterclockwise soliton output spectrum [1].

In a mode-locked fiber laser, when a stable soliton is generated, its state of polarization (SOP) transits to stable state. But if the soliton state in the cavity changes, the response of its SOP will be complicated. Klein et al. investigated the polarization dynamics in this unstable soliton state [2]. When excess pump power is injected into the laser cavity, the soliton formation process experiences relaxation oscillation, and the SOP of the soliton will change accordingly in this stage. At the same time, excessive power injection usually leads to the generation of multiple solitons. When two solitons generated in the cavity are close enough, and interact with each other, they can easily collide and merge into one soliton. Even though the two solitons are produced with different SOPs, as they pass each other, their SOPs move closer together, until the polarizations completely coincide when they collide (Figure 7.2).

When multiple solitons are closely combined to form a stable bound state, its SOP will also change to a stable state. The polarization evolution of bound states with tight and loose states was reported by Mou et al [3]. It is found by experiments that the SOP of the bound state, whether it is a tight or a loose state, is determined. By further pumping the current and tuning the PC, combined third-order harmonic mode locking with bound-state solitons is observed. This harmonic mode-locking operation changes the anisotropy (linear and circular birefringence) of the medium due to polarization hole burning, so the SOP of the pulse is located on a circle, as shown in Figure 7.3c.

## 7.2  POLARIZATION DYNAMICS OF NORMAL DISPERSION MODE-LOCKED LASERS

In the past decades, it has been found that the fiber lasers with net normal dispersion, could produce a kind of optical solitons, the dissipative soliton (DSs). They are remarkably localized structures that arise from the dynamical balance between nonlinearity, dispersion, and environmental energy exchange, which are highly coherent

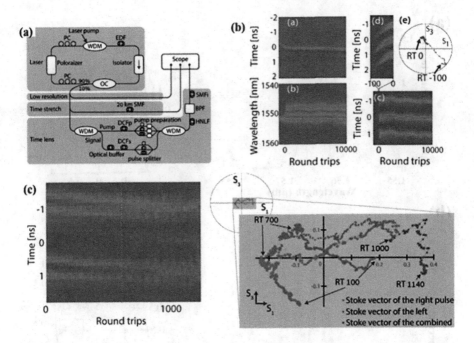

**FIGURE 7.2** (a) Experimental setup for measuring the temporal, spectral, and polarization dynamics of solitons in polarization-mode-locked fiber lasers. (b) Measured soliton dynamics showing relaxation oscillations and polarization evolution. (c) Time-lensing measurements showing the collision and merging of two solitons into a single soliton and polarization evolution [2].

solutions of nonlinear wave equations. The dynamics of the solitons formed in the normal dispersion fiber cavities are governed by the Ginzburg-Landau equation (GLE), which takes account not only of the cavity dispersion and nonlinearity but also of the gain and loss. Differing from solitons formed in the negative group velocity dispersion (GVD) regime, DSs formed in the large positive GVD regime have different soliton-shaping mechanism, and furthermore are strongly frequency chirped. Recently, the DSs have been extensively concerned owing to their high pulse energy, repetition rate, and improved output stability [4, 5]. Therefore, it would be worth investigating polarization dynamics in DSs, and determining the distinct features of the buildup process. It will provide a deeper insight into the research and application of dissipative soliton fiber lasers. At present, several polarization dynamics of normal dispersion cavity fiber lasers have been reported, whatever numerical simulation and experimental studies.

The dynamics of polarization evolution in DS fiber lasers were experimentally investigated by Gao et al [6]. The fiber ring laser cavity with net normal dispersion of 0.171 ps$^2$ is mode-locked by a saturable absorber (SA). The probabilistic polarization distributions of DSs in the system is described. In the experimental method, the filtered output optical spectra under different pump powers are detected by polarization state analyzer, and the phase diagram of the ellipticity angle and the spherical

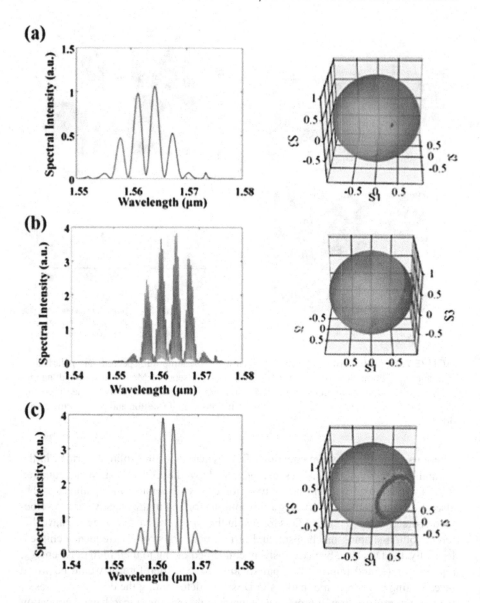

**FIGURE 7.3**   (a) Tightly bound polarization-locked vector bound-state solitons. (b) Loosely bound polarization-locked vector bound-state solitons. (c) Vector bound-state soliton with slow polarization evolution states for three-pulse harmonic mode-locking operation [3].

orientation angle is introduced to reveal the invariant polarization relationship between the wavelength components of the laser output field. Figure 7.4 illustrates the SOPs of different filtered wavelengths (a bandwidth with 0.2 nm) from the dissipative fiber laser cavity, for various levels of pump power. The corresponding SOPs for each wavelength are evolving from a random cloud into a fixed narrow domain as the pump power grows larger. When a stable DS is formed, a well-defined SOP

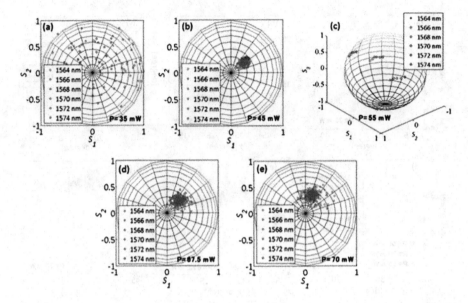

**FIGURE 7.4** (a) Evolution of SOPs for various filtered wavelengths and different pump powers. (b)–(e) Normalized Stokes parameters represented on the Poincaré sphere for pump powers of 35, 45, 55, 67.5, and 70 mW, respectively [6].

trajectory versus frequency is observed on the Poincaré sphere. However, when unstable DSs is produced, their SOPs exhibit fluctuation. When the pump power is stronger, these intense fluctuations trend diffuses more on the Poincaré sphere. The instability and fluctuations of SOPs on the two edges of the spectrum reveal a new aspect of the complex dynamics in DS formation.

The phase diagram (ellipticity angle $\chi$ and the spherical orientation angle $\psi$) is used to describe the characteristics of the SOP distribution.

$$\begin{cases} \chi = \dfrac{1}{2}\arctan\left(\dfrac{S_3}{\sqrt{S_1^2 + S_2^2}}\right) \\[4mm] \psi = \dfrac{1}{2}\arctan\left(\dfrac{S_2}{S_1}\right) \end{cases} \tag{7.1}$$

Here, $S_1$, $S_2$, and $S_3$ are the Stokes parameters of the laser SOP. By introducing phase diagram parameters, as shown in Figure 7.5, it is found that the SOP in the central region of the DSs evolves with frequency along a smooth trajectory on the Poincaré sphere, whereas SOPs cross the two edges of the spectrum evolve with frequency in different directions with respect to the evolution of the central spectrum. When the spectrum is broadened, the SOP of the newly generated frequencies will be randomly dispersed. Further increasing of the pump power result in DS deterioration, and lead to the observation of soliton explosions. And a new type of optical rogue waves (RWs) is confirmed in polarization domain, namely, polarization rogue waves (PRWs).

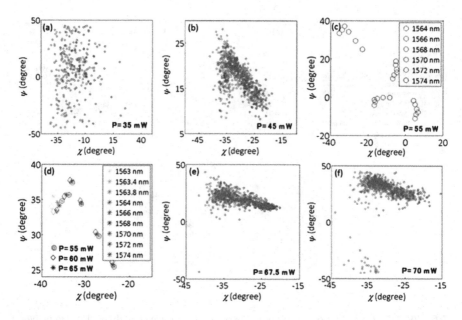

**FIGURE 7.5** (a)–(c), (e), (f) Phase diagrams based on the ellipticity angle $\chi$ and spherical orientation angle $\psi$, corresponding to the SOPs of various filtered wavelengths and different pump powers in Figure 7.4(a)–(e). (d) Phase diagram for a stable DS, where the pump power is either 55 mW (circled dots), 60 mW (rectangle dots), or 65 mW (star dots), respectively. The SOPs on the edges of the spectrum exhibit fluctuations [6].

The polarization dynamics under different output coupling ratios in DS lasers mode-locked by NPR were numerically studied by Kong et al [7]. The scheme diagram is shown in Figure 7.6a. The unidirectional operation of the ring cavity is enforced by the inline isolator. The output is extracted from the PBS ejection port. Numerical studies on pulse propagation in fiber sections are based on a coupled GLE model for the orthogonal electric field SOPs [8]. Figures 7.6b–e show the polarization dynamics at different intracavity locations. The points on the surface of the Poincaré sphere represent the SOPs at every time slice. It is observed that the SOPs vary with time across the pulse, and the SOPs of the pulse peak before the polarizer depend on the output coupling ratio: under high output coupling, they are nearly circular polarizations; but in the medium and low output coupling case, they are nearly linear polarizations.

The dissipative vector solitons (DVSs) in a dispersion-managed cavity fiber laser with large net positive cavity GVD were experimentally observed by Zhang, and theoretically predicted in a fiber laser mode-locked with a semiconductor saturable absorber mirror (SESAM) [9]. This fiber laser has a typical dispersion-managed cavity with a net normal dispersion of about $0.027\text{ps}^2$. A rotatable external cavity polarizer is used to highlight the vector nature of soliton. The polarization rotation of the soliton can be easily identified by oscilloscope trace measurement. The polarization of the soliton is observed to rotate along the cavity by comparing the intensity with or without an external polarizer, as shown in Figure 7.7b. In the experiment,

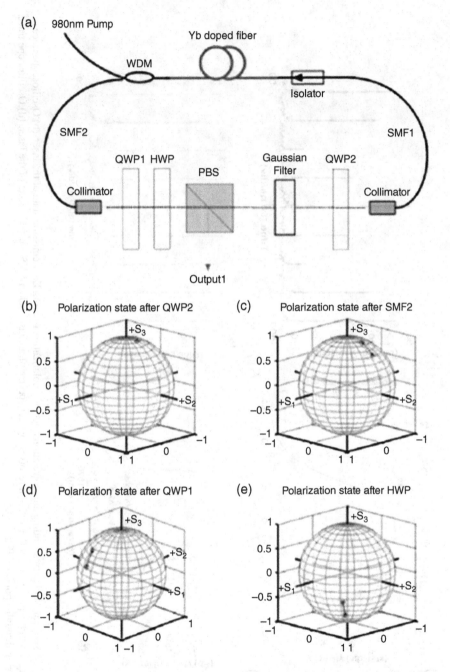

**FIGURE 7.6** (a) Schematic diagram of NPR mode-locked dissipative soliton fiber laser. (b)–(e) SOPs across the pulse at different intracavity locations in high output coupling case [7].

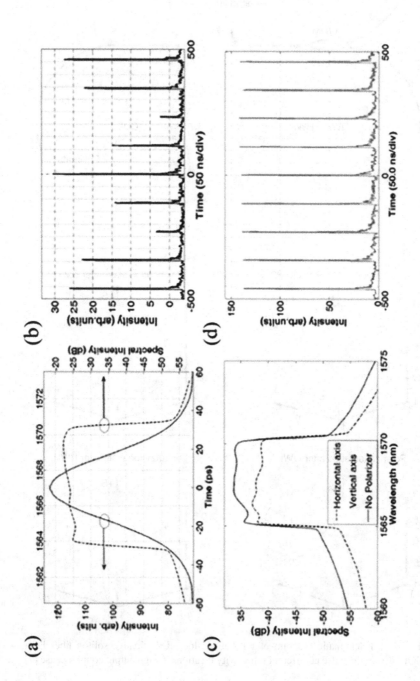

**FIGURE 7.7**  (a) Spectrum and corresponding autocorrelation trace of a polarization rotating DVS emission state of the laser. (b) Oscilloscope trace of (a) after passing through a polarizer. (c) Polarization resolved optical spectra of a phase locked DVS emission state of the laser. (d) Oscilloscope trace of (c) after passing through a polarizer [9].

polarization locked vector solitons can be obtained by controlling the linear cavity birefringence. It has the characteristic fixed polarization during circulation in the laser cavity. In addition, the GLE is used to describe the pulse propagation in the weakly birefringent fibers in the simulations, and they are found in agreement of the experimental observations.

## 7.3   POLARIZATION DYNAMICS OF NEAR-ZERO NET DISPERSION MODE-LOCKED LASERS

In particular, when the net dispersion is close to zero, and the normal and anomalous dispersion fibers coexist in the laser cavity, the operation states of lasers and interaction processes of pulses will be controlled by the weak fiber birefringence induced by the polarization controller or any other polarization-dependent devices in non-polarization-maintaining systems. Benefitting from this new dimensional controlling, telecommunication capacity will be further enhanced by multiplexing of SOP [10], manipulation of vector magnetization can also be achieved by using polarized laser [11], various dynamic patterns ultra-precision machining can be realized by the polarization adjustment of laser processing sources [12]. Therefore, it is extremely important to study the polarization characteristics and the dynamic interaction processes of ultrafast lasers with the near-zero net dispersion, which will help to deepen understanding of the evolution dynamics of vector pulses, and provide a new insight for designing and controlling the mode-locked laser with polarization dependent output pulses.

Recently, Cao, et al have investigated novel polarization-dependent pulsing procedures in ultrafast fiber laser cavity [13]. Utilizing polarization controller and dispersion management, the weak birefringence effect in fiber means that the pulse propagating in normal and abnormal dispersion mediums enables ultra-short pulses strong polarization dependence and energy coupling. Their experimental system and measurement results are shown in Figure 7.8. Under the action of double-scroll chaotic polarization attractors in the ultrafast fiber laser [14], the energy of the main pulse is dispersed into the respective SOPs when the pulse changes from the fixed-point SOP to the semicircular scattered SOP on the Poincaré sphere. The gain of the main pulse will decrease gradually due to the polarization burn hole in the gain medium, but the number of inverted particles in the gain medium is not exhausted, so the dispersion wave in other SOPs in the laser cavity will be amplified, and the secondary pulse will be generated. In a near-zero dispersion laser cavity, the pulse is stretched between the normal dispersion erbium-doped fiber (EDF) and the anomalous dispersion single mode fiber (SMF), so thousands of roundtrips are required to reach the condensed phase. Finally, when the two pulses reach the condensing stage, their energy begins to stabilize.

The polarization attractor of vector soliton was proposed by Sergey in 2014. Based on the coupled Schrodinger equation and Ginzburg-Landau equation, combined with the Grassberger-Procaccia algorithm, the researchers theoretically calculated another strange laser polarization attractor different from Lorenz, Rossler and Ikeda attractors, as shown in Figures 7.9a, b. The SOP of the output ultra-fast laser evolves linearly or nonlinearly between two definite points on the Poincaré sphere.

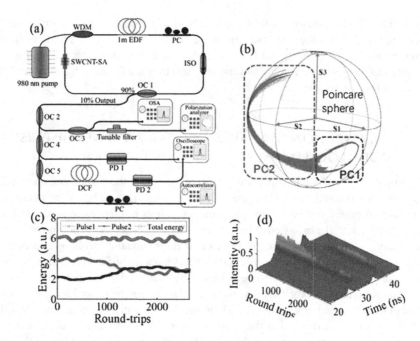

**FIGURE 7.8**   (a) Structure diagram of the near-zero dispersion ultrafast laser and measurement system. (b) All the SOPs of the output pulses. (c) Spectral energy evolution of total output, pulse 1 and pulse 2 under transition state of PC1 region switching to PC2 region. (d) Energy coupling between two pulses measured by dispersion Fourier transform technique [13].

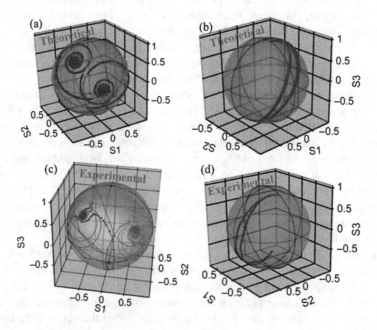

**FIGURE 7.9**   (a), (b) Stokes parameters on the Poincaré sphere obtained by experiment. (c), (d) Stokes parameters on the Poincaré sphere obtained by simulation [14].

Experimentally, the polarization controller and pump power are combined to control the dipoles generated by absorption and emission in carbon nanotube (CNT) and EDF of the ultrafast fiber laser cavity; thus, the evolutionary dynamics of laser polarization as shown in Figures 7.9c, d are demonstrated.

## 7.4  POLARIZATION DYNAMICS OF PARTIALLY MODE-LOCKED FIBER LASERS

Fiber laser cavity not only establishes aforementioned high-coherent stable mode-locked pulses, but also generates unstable partially mode-locked (PML) pulses under extraordinary conditions [15, 16]. By excessively pumping the fiber cavity, complex and randomly fluctuating regime occurs. Spectrum with giant envelope and irregular intensity distribution of temporal pulse trains are typical characteristics. Due to similarity to noise-like bursts, outputs for PML have also been named as *noise-like pulses*. Meanwhile, a sharp coherent peak, together with apparent noise base, can be observed in autocorrelation trace. In addition, abundant polarization dynamics have been unveiled in PML fiber lasers in these recent studies.

Kbashi et al. found that the change of the intra-cavity SOP leads to the formation of chaotic pulses in the PML regime [17]. In this case, chaotic mode-locked pulses undergo the impact of vector-resonance-multimode instability (VRMI). Due to the spatial modulation of intra-cavity SOP, various spatio-temporal structures with additional satellite frequencies in spectral domain under VRMI appear. The effect of satellites and fundamental frequencies is responsible for the emergence of quasi-periodic motion and chaotic behavior. As shown in Figure 7.10a, torus trajectory at the Poincaré sphere corresponding to quasi-periodic motion can be observed. Formation of chaos originates from the destruction of the torus. According to the temporal variation (Figures 7.10b, c), the evolution of polarization dynamics is slow. It is noted that, optical RWs accompany chaotic behavior in the PML regime [17–19].

Furthermore, the evolving SOP for filtered wavelengths in PML at different pump powers has been investigated by Gao et al [20, 21]. Figure 7.11 depicts that the corresponding SOPs for wavelengths far away from the center bifurcate into a cross-like shape on the Poincaré sphere. Perpendicular lines result from the multiple wave-mixing processes, giving rise to new frequencies with independently output SOP on the Poincaré sphere. SOP turbulence is accompanied by the loss of system coherence. When the laser cavity is pumped exceeding 250 mW, irregular SOPs located outside of the main polarization directions occur. The higher intracavity power, the more scattered distribution of SOPs. This results from a cascade of successive period-doubling bifurcations. When the pump power is larger than 600 mW, the cascaded four-wave-mixing (FWM) leads to a fully developed turbulent evolution, and random SOPs of filtered wavelengths appears.

In addition to the optical RWs in PML, the irregular SOP is associated with the emergence of a new kind of RWs, viz. polarization rogue waves (PRWs). Due to the vector feature of PRWs, the characterization method is introduced with the relative distance, r, between two SOPs. The probabilities of the relative distance between various SOPs under different pump powers for filtered wavelengths are depicted in Figure 7.12. During the laminar-turbulence transition of the SOP, abnormal L-shaped probability

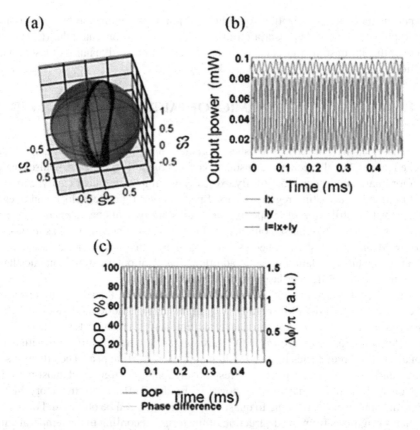

**FIGURE 7.10**  Polarization laser dynamics of PML chaotic solitons. (a) Trajectories on normalized Poincaré sphere. (b) The output powers versus time. (c) Degree of polarization and the phase difference versus time [17].

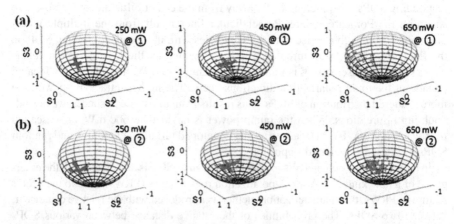

**FIGURE 7.11**  Measured SOP for filtered wavelengths under various pump powers in PML regime. (a)–(e) for 1547.6 nm, 1556.1 nm, 1,565 nm, 1574.1 nm and 1582.7 nm, respectively [20].

(*Continued*)

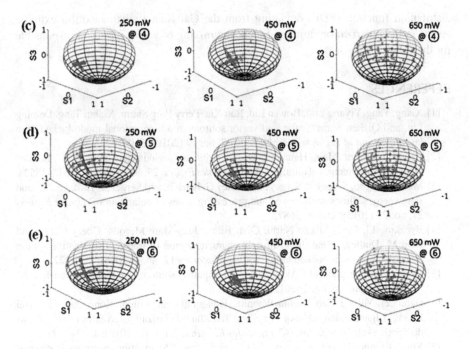

**FIGURE 7.11**   (Continued)

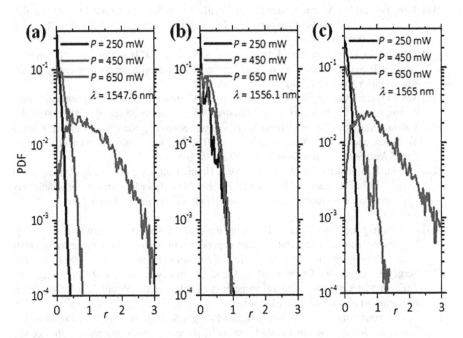

**FIGURE 7.12**   Logarithm of the PDFs of relative distance between various points on Poincaré sphere. (a)–(e) for 1547.6 nm, 1556.1 nm and 1565 nm, respectively [21].

distribution function (PDF) deviating from the Gaussian statistics exhibit extreme events in the polarization domain. Stochastic mixing of vector modes is responsible for the PRWs.

## REFERENCES

[1] Xiang, Yang, Yiyang Luo, Bowen Liu, Ran Xia, Perry Ping Shum, Xiahui Tang, Deming Liu, and Qizhen Sun. "Scalar and vector solitons in a bidirectional mode-locked fibre laser." *Journal of Lightwave Technology* 37, no. 19 (2019): 5108–5114.

[2] Klein, Avi, Sara Meir, Hamootal Duadi, Arjunan Govindarajan, and Moti Fridman. "Polarization dynamics of ultrafast solitons." *Optics Express* 29, no. 12 (2021): 18512–18522.

[3] Mou, Chengbo, Sergey V. Sergeyev, Aleksey G. Rozhin, and Sergei K. Turitsyn. "Bound state vector solitons with locked and precessing states of polarization." *Optics Express* 21, no. 22 (2013): 26868–26875.

[4] Ryczkowski, Piotr, Mikko Närhi, Cyril Billet, Jean-Marc Merolla, Göery Genty, and John M. Dudley. "Real-time full-field characterization of transient dissipative soliton dynamics in a mode-locked laser." *Nature Photonics* 12, no. 4 (2018): 221–227.

[5] Grelu, Philippe and Nail Akhmediev. "Dissipative solitons for mode-locked lasers." *Nature Photonics* 6, no. 2 (2012): 84–92.

[6] Gao, Lei, Yulong Cao, Stefan Wabnitz, Hongqing Ran, Lingdi Kong, Yujia Li, Wei Huang, Ligang Huang, Danqi Feng, and Tao Zhu. "Polarization evolution dynamics of dissipative soliton fiber lasers." *Photonics Research* 7, no. 11 (2019): 1331–1339.

[7] Kong, Lingjie, Xiaosheng Xiao, and Changxi Yang. "Polarization dynamics in dissipative soliton fiber lasers mode-locked by nonlinear polarization rotation." *Optics Express* 19, no. 19 (2011): 18339–18344.

[8] Tsoy, Eduard N., Adrian Ankiewicz, and Nail Akhmediev. "Dynamical models for dissipative localized waves of the complex Ginzburg-Landau equation." *Physical Review E* 73, no. 3 (2006): 036621.

[9] Zhang, Han, Dingyuan Tang, Luming Zhao, Xuan Wu, and Hwa-Yaw Tam. "Dissipative vector solitons in a dispersion-managed cavity fiber laser with net positive cavity dispersion." *Optics Express* 17, no. 2 (2009): 455–460.

[10] VanWiggeren, Gregory D., and Rajarshi Roy. "Communication with dynamically fluctuating states of light polarization." *Physical Review Letters* 88, no. 9 (2002): 097903.

[11] Kanda, Natsuki, Takuya Higuchi, Hirokatsu Shimizu, Kuniaki Konishi, Kosuke Yoshioka, and Makoto Kuwata-Gonokami. "The vectorial control of magnetization by light." *Nature Communications* 2, no. 1 (2011): 1–5.

[12] Öktem, Bülent, Ihor Pavlov, Serim Ilday, Hamit Kalaycıoğlu, Andrey Rybak, Seydi Yavaş, Mutlu Erdoğan, and F. Ömer Ilday. "Nonlinear laser lithography for indefinitely large-area nanostructuring with femtosecond pulses." *Nature Photonics* 7, no. 11 (2013): 897–901.

[13] Cao, Yulong, Lei Gao, Yujia Li, Hongqing Ran, Lingdi Kong, Qiang Wu, Ligang Huang, Wei Huang, and Tao Zhu. "Polarization-dependent pulse dynamics of mode-locked fiber laser with near-zero net dispersion." *Applied Physics Express* 12, no. 11 (2019): 112001.

[14] Sergeyev, Sergey V., Chengbo Mou, Elena G. Turitsyna, Alexey Rozhin, Sergei K. Turitsyn, and Keith Blow. "Spiral attractor created by vector solitons." *Light: Science & Applications* 3, no. 1 (2014): e131–e131.

[15] Kwon, Youngchul, Luis Alonso Vazquez-Zuniga, Seungjong Lee, Hyuntai Kim, and Yoonchan Jeong. "Numerical study on multi-pulse dynamics and shot-to-shot coherence property in quasi-mode-locked regimes of a highly-pumped anomalous dispersion fiber ring cavity." *Optics Express* 25, no. 4 (2017): 4456–4469.

[16] Lee, Seungjong, Luis Alonso Vazquez-Zuniga, Hyuntai Kim, Youngchul Kwon, Kyoungyoon Park, Hansol Kim, and Yoonchan Jeong. "Experimental spatio-temporal analysis on the shot-to-shot coherence and wave-packet formation in quasi-mode-locked regimes in an anomalous dispersion fiber ring cavity." *Optics Express* 25, no. 23 (2017): 28385–28397.

[17] Kbashi, Hani J, Marina Zajnulina, Amos G Martinez, and Sergey V. Sergeyev. "Mulitiscale spatiotemporal structures in mode-locked fiber lasers." *Laser Physics Letters* 17, no. 3 (2020): 035103.

[18] Lecaplain, Caroline and Philippe Grelu. "Rogue waves among noiselike-pulse laser emission: An experimental investigation." *Physical Review A* 90, no. 1 (2014): 013805.

[19] Pottiez, Olivier, Hugo Enrique Ibarra-Villalon, Yazmin Esmeralda Bracamontes-Rodríguez, Josué A. Minguela-Gallardo, Ernesto García-Sanchez, Jesús Pablo Lauterio-Cruz, Juan Carlos Hernandez-Garcia, Miguel A. Bello-Jiménez, and Evgeny A. Kuzin. "Soliton formation from a noise-like pulse during extreme events in a fibre ring laser." *Laser Physics Letters* 14, no. 10 (2017): 105101.

[20] Gao, Lei, Tao Zhu, Stefan Wabnitz, Min Liu, and Wei Huang. "Coherence loss of partially mode-locked fibre laser." *Scientific Reports* 6, no. 1 (2016): 1–12.

[21] Gao, Lei, Tao Zhu, Stefan Wabnitz, Yujia Li, Xiaosheng Tang, and Yulong Cao. "Optical puff mediated laminar-turbulent polarization transition." *Optics Express* 26, no. 5 (2018): 6103–6113.

# 8 Dual-Wavelength Fiber Laser for 5G and Lidar Applications

*Hani J. Kbashi and Vishal Sharma*

## CONTENTS

8.1 Introduction ................................................................................................205
8.2 Dual-Wavelength Fiber Laser Experimental Setup and Characterization ....207
8.3 DWFL for Millimeter Waves (5g) Transmission Applications ...................211
    8.3.1 Transmission of mmW Waves over Radio-Over-Fiber (ROF) Link ........................................................................................................211
    8.3.2 Transmission of mmW Waves over a Free-Space Optics (FSO) Link ........................................................................................................213
8.4 DWFL for Lidar Applications .....................................................................217
    8.4.1 Experimental Results and Discussion .............................................217
8.5 Chapter Conclusions ....................................................................................220
Acknowledgments ................................................................................................220
References .............................................................................................................220

## 8.1 INTRODUCTION

Recently, fiber laser as a photonic source has attained revolutionary evolution in several fundamental technologies and applied sciences areas, including natural science studies, microwave photonics, photonics-based telecommunication, optical spectroscopy, material processing, meteorology, sensing, surveillance & navigation, and medical-related applications (Dennis et al., 1997; Udem et al., 2002; Fortier et al., 2011; Sibbett et al., 2012; Kolpakov et al., 2016; Yadav et al., 2017; Kbashi et al., 2018; Tsai et al., 2018). Moreover, fiber lasers as a source of millimeter-wave/Terahertz (mmW/THz) wave generation (Kbashi, 2021a; Kbashi, 2021b) for telecommunication-related applications has gained popularity in the past few years through the interplay of gain/loss mechanisms along with linear and nonlinear effects. By maintaining these factors in a controlled manner, tuneable transmitters in mmW/THz band with high flexibility and stability are attainable. Additionally, a tuneable dual-wavelength fiber-laser (DWFL) for the generation of mmW is gaining considerable attraction nowadays due to its vast applications in the research and manufacturing sectors, including autonomous vehicle-related industry and meteorology. Furthermore, compared to other various techniques (Adany et al., 2009; Gao et al., 2012; Morris et al., 2012), the significant interest in dual-wavelength

DOI: 10.1201/9781003206767-8

generation using a single fiber laser offers a high mutual coherence, common noise cancellation, compact size of a high-frequency signal source together with a simple and economical design along with high scalability to the state-of-the-art microwave-photonics networks. Also, due to low phase-noise and high phase-coherence, DWFL lasers have the potential for the realization/implementation of futuristic 5G/6G networks, for instance, radio over fiber (RoF) systems (Kbashi, 2021b), free-space optics (FSO) (Kbashi, 2021c), hybrid microwave-optics systems (Zhu and Li, 2017), photonics-based radar systems (Sharma et al., 2021), and terahertz (THz) radiation-related applications (Hu et al., 2018).

Alternatively, the state-of-the-art Autonomous Vehicles (AVs) demand high-resolution radar imaging to track multiple targets in complex traffic scenarios for all-weather situations. Subsequently, an mmW band (74 GHz–77 GHz) is preferred to achieve a high bandwidth of ≈4 GHz for the manufacturers. According to the range-resolution equation, $L_{res} = c/2B$, the broadened signals improve the radar resolution and reduce its size considerably (Zhang et al., 2017; Ramasubramanian and Ramaiah, 2018; Sharma and Sergeyev, 2020). Additionally, the environmental fluctuations become catastrophic in the urban- and industry-predominated zones due to dense gasses, massive smoke, and airborne particles. It becomes difficult to retrieve the range-speed measurements unambiguously from the weak echoes. So, it becomes a challenging task to identify the mobile/immobile targets of different dimensions on the road accurately.

The enhanced radar resolution further augments the detection capability and can play a crucial role in situations of poor visibility. In addition, the current AVs are equipped with the driver assistance system (DAS), including the conventional microwave radar systems, three-dimensional cameras, and signal processing units. However, it offers a limited detection range with marginal resolutions at high power requirements (Sharma and Sergeyev, 2020). Therefore, the light-based automotive radar system proves to be an attractive candidate for state-of-the-art surveillance, tracking, and navigation-like applications.

Alternatively, the high-power and phase-stable mmW generation in the preferable frequency band (74 GHz–77 GHz) using the traditional electronics-based techniques is a thought-provoking task. Thus, dual-wavelength fiber lasers have been designed efficiently in the last few years to realize self-driving and surveillance-related applications in this preferred band (Adany et al., 2009; Gao et al., 2012; Khattak and Wei, 2018). Moreover, *DWFL*-based mmW generation is free from using a microwave source, thus offering less complexity, low phase noise, and being economical. However, the erbium-doped fiber-employed DWFL lasers experience mode-competition problems in the 1550 nm range. Several expensive and complex demonstrations to manage this mode-competition issue have been established (Zhou et al., 2015; Lian et al., 2017; Yan et al., 2017; Khattak and Wei, 2018). Subsequently, this chapter demonstrates an erbium-doped fiber-based tunable DWFL to generate mmW at 77 GHz in a simple configuration without using any additional high-quality microwave source. These generated mmW waves are utilized to develop a linear photonics-based radar system and are investigated under severe atmospheric perceptions. The designed DWFL is tunable over a range of 0.1–0.9 nm, corresponding to the mmW frequency range of ~12.3–110 GHz with a tunable step of ~10 GHz. To validate the proposed fiber laser in 5G-related applications, successful transmission of

16-QAM modulated signals at a data rate of 10 Gb/s over the RoF and FSO link is demonstrated. Further, the proposed laser is developed to realize an automotive photonics-based radar system in successive sections.

## 8.2 DUAL-WAVELENGTH FIBER LASER EXPERIMENTAL SETUP AND CHARACTERIZATION

The experimentally simple and compact DWFL configuration is illustrated in Figure 8.1a. A nonlinear polarization rotation (NPR) ring-cavity is formed consisting of one-meter Er-doped fiber, 10 m of polarization maintaining fiber, and 9.1 m of single-mode slandered telecommunication fiber (SMF), amounting to an overall cavity length of 20.1 m. The high concentration erbium-doped active fiber (Liekki Er80-8/125) is pumped using a 980-nm laser diode via a wavelength-division-multiplexer (WDM-980/1,550 nm). While the two polarization controllers (PCs), together with the polarization-independent optical isolator (ISO), achieve a phase-stable NPR lasing action and adjusting/maintaining the laser net cavity-birefringence; ISO also ensures a unidirectional circulation inside the cavity. An accurate optical filtering inside the laser cavity, over a wide spectral band to attain a uniform channel-spacing, is attained by incorporating a 10-m high birefringence (HiBi) fiber (NA = 0.125; core/Cladding = 8.5/125 µm). It helps in generating high phase-stable dual wavelengths by stabilizing the state of polarization. A 90:10 fused fiber output coupler is incorporated to feedback about 90% of the signal power inside the cavity and about 10% of the signal power is used for the spectral and temporal study using a fast 17 GHz photodetector (InGaAsUDP-15-IR-2_FC) linked to a 2.5 GHz sampling oscilloscope (Tektronix DPO7254). A polarimeter (Thorlabs IPM5300) of 1 µs resolution in a time interval ≈1 ms (25–40,000 roundtrips) is used to compute the normalized

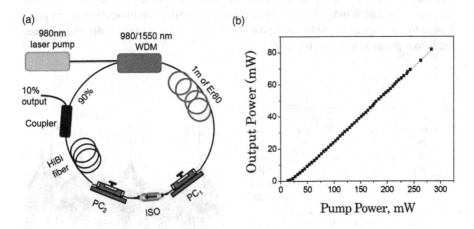

**FIGURE 8.1** (a) The experimental setup of the proposed switchable dual-wavelength fiber laser: Er80, High concentration erbium-doped fiber; $PC_{1,2}$, Polarization controllers; WDM, A wavelength division multiplexing coupler; ISO, An optical isolator; HiBi, High birefringence fiber. (b) The measured output power from the 10% of the output coupler vs pump power (Kbashi, 2021). (Adapted from Kbashi et al., 2021a.)

Stokes parameters ($S_1$, $S_2$, $S_3$), and degree of polarization (*DOP*). An optical spectrum analyzer (Yokogawa AQ6317B) of maximum resolution ≈20 pm, and a radio frequency spectrum analyzer observe the optical and the RF spectrum respectively. Figure 8.1b shows the measured output power from 10% of the output coupler vs pump power, showing a very low threshold of laser ≈20 mW and a high output power ≈80 mW with a power efficiency of ≈35%.

By proper adjustment of the pump power initially at 20 mW and the net birefringence of the cavity using the two PCs, a dual-wavelength lasing is attained with a wavelength spacing of 0.1 nm and then, it is perfectly tuned from 0.1–0.89 nm sequentially, as shown in Figure 8.2a, conforming to beat frequency ≈12.5 GHz–110 GHz (mmW), as shown in Figure 8.2b. The observations show that initially, the dual-wavelength peaks at 0.1 nm are located at 1560.0 nm and 1560.1 nm with a narrow linewidth of 0.125 nm and 0.13 nm, respectively. However, the laser peaks at 0.89 nm are observed at 1559.54 nm and 1560.43 nm with a narrow linewidth of 0.12 nm and 0.1 nm, respectively.

Further, the phase-stability of the generated dual-wavelength peaks at 1559.5 nm and 1560.35 with a wavelength spacing of 0.85 nm are observed by analyzing the output spectra for 60 minutes with a scanning-interval of 10 minutes, as illustrated in Figure 8.3a. Figure 8.3a shows phase-stable spectra, free from any wavelength shift, during the observation period with an optical signal-to-noise ratio (≈40 dB). Further, a low power fluctuation of 0.32 dB and wavelength peak-fluctuation ≈0.03 nm at 56 GHz are also measured over the observation period (Kbashi, 2021). A phase-stable dual-wavelength operation may attain due to maintaining the state of polarization by incorporating the HiBi fiber and it may efficiently overpower the mode competition of the homogenous line-broadening and cross-gain saturation in the Er-doped fiber. Moreover, the PCs with HiBi may provide a wavelength-dependent polarization rotation and fluctuates in the state of the polarization (SOPs) across the multiple wavelengths. Together, it may also support attaining an optimal amplification at low pump power by managing the SOPs and providing a linearly polarized output that further leads to a phase-stable tunable laser.

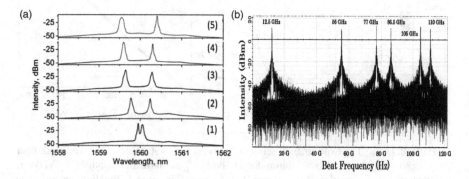

**FIGURE 8.2**   (a) Tunable dual-wavelength emission spacing at (1) 0.1 nm; (2) 0.46 nm; (3) 0.63 nm; (4) 0.7 nm; and (5) 0.89 nm. (b) The corresponding mmW from 12.5 GHz to 110 GHz (Kbashi, 2021). (Adapted from Kbashi et al., 2021a.)

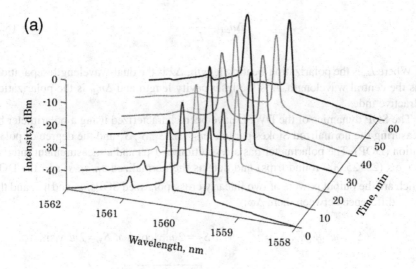

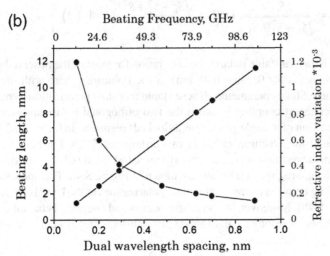

**FIGURE 8.3** (a) Stability spectra of the DWFL-based generation with wavelength-spacing of 0.85 nm. (b) The beat-length and the two orthogonal polarization refractive-indices variation as a function of different dual-wavelength spacing. (Adapted from Kbashi et al., 2021a).

The beat length ($L_b$) and the difference between the two orthogonal polarization refractive indices as a function of the dual-wavelength spacing are calculated using Equations 8.1 and 8.2. The results are shown in Figure 8.3b. The beating length is tuned from ≈1.3 mm to ≈12 mm by maintaining the total cavity birefringence via controlled adjustment of the PCs. It leads to an increase in the free spectral range of the HiBi fiber and measures the difference between the two orthogonal polarization refractive indices in the range of minima to maxima as $1.35 \times 10^{-4}$ to $1.2 \times 10^{-3}$ in this experiment.

$$L_b = \frac{\Delta\lambda}{\lambda} \cdot L \qquad\qquad (8.1)$$

$$\Delta n_{\text{eff}} = \frac{\lambda}{L_b} \qquad (8.2)$$

Where $L_b$ is the polarization-beating length, $\Delta\lambda$ is the dual-wavelength separation, $\lambda$ is the central wavelength, $L$ is the laser cavity length and $\Delta n_{\text{eff}}$ is the polarization refractive index.

The SOP dynamics of the DWFL have been characterized using a polarimeter by measuring the normalized Stokes parameters, i.e., $S_1$, $S_2$, $S_3$, and the degree of polarization (DOP). The polarimeter has a resolution of 1 μs and a measurement interval of 1 ms (25–25,000 round trips) and detects the normalized $s_1$, $s_2$, $s_3$ and the DOP which are the output powers of two linearly cross-polarized SOPs, $|u|^2$ and $|v|^2$, and the phase difference between them $\Delta\varphi$:

$$S_0 = |u|^2 + |v|^2, \ S_1 = |u|^2 - |v|^2, \ S_2 = 2|u||v|\cos\Delta\varphi, \ S_3 = 2|u||v|\sin\Delta\varphi,$$

$$S_i = \frac{S_i}{\sqrt{S_1^2 + S_2^2 + S_3^2}}, \ DOP = \frac{\sqrt{S_1^2 + S_2^2 + S_3^2}}{S_0}, (i = 1,2,3) \qquad (8.3)$$

Figure 8.4 illustrates a stable polarization operation for most of the observed dual-wavelength regimes (as for 0.1 nm–0.48 nm) in the Poincaré sphere with the axis defined by the three Stokes parameters. These stable and slow polarization dynamics indicate a high coherent coupling between the two orthogonal polarizations which leads to the production of a single-pulse (mode-locked) regime as in 0.1 nm, 0.29 nm, and 0.48 nm due to the synchronization of two orthogonal SOPs. Furthermore, the laser system starts desynchronizing the two orthogonal SOPs at 0.7 nm while attaining a complex chaotic regime at 0.89 nm, as shown in Figure 8.4a. This indicates an incoherent coupling between the two orthogonal polarizations as the DOP is decreased to <40% (Figure 8.4b). Moreover, this synchronization and coupling behavior can be described via the vector resonance multimode instability (Sergeyev et al., 2017).

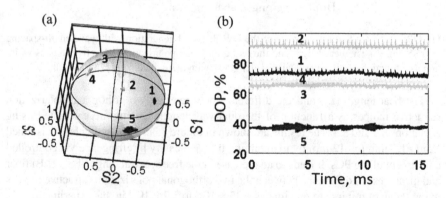

**FIGURE 8.4** (a) The SOP. (b) DOP of the switchable dual-wavelength fiber laser-based generation at different wavelength spacing, (1), 0.1 nm; (2), 0.29 nm; (3), 0.48 nm; (4), 0.7 nm; and (5), 0.89 nm. (Adapted from Kbashi et al., 2021a).

## 8.3 DWFL FOR MILLIMETER WAVES (5G) TRANSMISSION APPLICATIONS

### 8.3.1 TRANSMISSION OF mmW WAVES OVER RADIO-OVER-FIBER (ROF) LINK

The proposed tunable DWFL designed experimentally in the previous section has enormous opportunities in terms of futuristic 5G/6G transmission systems, including RoF, FSO transmission systems, and automotive LiDAR systems. This section deals with the implementation of orthogonal frequency-division multiplexing (OFDM)-incorporated RoF transmission system to transmit a data rate of 10 Gb/s (16-QAM signals) over four different RF carrier frequencies, i.e., 56.7, 77.6, 86.2, and 110 GHz, which are generated by using the tunable DWFL laser, as shown in Figure 8.5. The OFDM-RoF system is demonstrated by co-simulating the *Optisystem*™ photonic module and *MATLAB*™ software. The antenna module is designed using the *MATLAB*™ software which is integrated with *Optisystem*™ using *MATLAB.dll* files. By co-simulating both software-based modeled sub-systems, the tunable DWFL-based mmW waves are generated and then transmitted over the OFDM-RoF link. Figure 8.5 shows the simulation setup for transmitting the proposed laser-based generated RF signals over the RoF link.

The dual-wavelength optical signals, for instance, 0.46 nm (~56.7 GHz) are separated using DWDM (1:2 splitter) into two wavelengths (1558.94 nm and 1559.40 nm). In the first arm of the splitter, the OFDM modulated data signals that are generated employing a 16-QAM digital modulation via a local oscillator (LO) are optically modulated over the optical frequency carrier of 1559.40 nm using a Mach Zehnder modulator (MZM). A suitable optical bandpass filter (OBPF) has been used with a center wavelength of 1559.40 nm to filter out the high sideband and only obtain the lower sideband of the QAM signals i.e., LSB-QAM. This LSB-QAM signal is applied to another DWDM (2:1 combiner) to combine with the unmodulated output signal in the second arm of DWDM. The obtained LSB-QAM modulated signals at the output of the combiner are propagated over SMF fiber (20 Km) and then intensified using a low-noise EDFA amplifier to restore the optical power budget. These modulated OFDM signals are mixed with the unmodulated carrier signals at the radio access unit (RAU) in a PIN photo-detector with a responsivity of 0.8 A/W to generate the required mmW signals, which are further then transmitted over a wireless link using a transmitting antenna (Gain ≈25 dBi). On the receiver side, the original information is retrieved by the demodulation process using the local oscillator (LO) and a microwave demodulator.

However, an additive white Gaussian noise (AWGN) wireless channel is designed using *MATLAB*™ and is integrated with the *Optisystem*™ photonic-module to realize the RoF system. In the 55–60 GHz frequency band, the atmospheric factors, such as absorption by atmospheric gasses, water vapor density, and other atmospheric constituents, offer a significant signal-fading under dry and standard atmospheric situations (Zibar et al., 2011). It causes a weak signal reception at the radio unit (RU) and leads to a short transmission range. Therefore, due to the high attenuation of signals in this frequency band with geometric losses, it is essential to restore the required signal power by applying a suitable amplification. Therefore, the received signals are amplified at the RU unit using a low-noise amplifier (LNA) of 17 dB in the

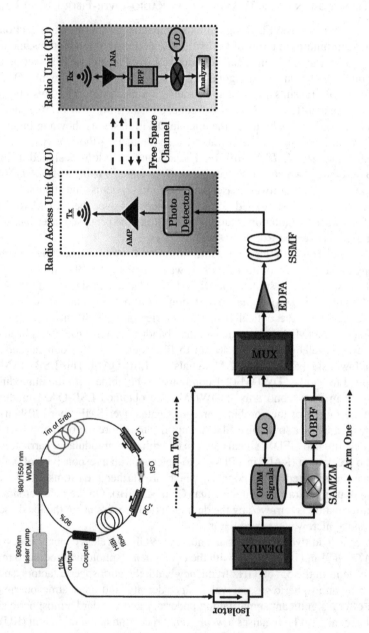

**FIGURE 8.5**  Simulation setup of Tunable DWFL-driven RoF transmission system. (Adapted from Kbashi et al., 2021b.)

demonstrated mmW band (56 GHz–110 GHz) with a noise figure of ≈5 dB (Bessemoulin et al., 2005; Zibar et al., 2011) after propagating through the wireless link. These amplified signals are applied to a band pass filter (BPF) after OFDM-demodulation to retrieve the transmitted data signals.

The performance evaluation of the DWFL-derived OFDM-RoF system is computed in terms of eye diagrams, bit-error-rate versus signal-to-noise ratio, and constellation diagram, as shown in Figures 8.6–8.8, which confirmed the successful transmission of OFDM data signals centered at the generated mmW frequencies (56.7, 77.6, 86.2, and 110 GHz) analogous to the dual-wavelength separation of the proposed dual-wavelength laser. Figure 8.6 shows the eye diagram estimation of the transmitted signals over SSMF of 20 Km at varied wavelength spacing at the radio access unit after photo-detection. Furthermore, the BER estimation as a function of wireless link length and SNR ratio is carried out to achieve the threshold BER of $10^{-3}$. The outcomes reveal that as the transmission occurs at a higher frequency of the millimeter-wave band, the SNR penalty increases to obtain the threshold BER ($10^{-3}$). As per the atmospheric impact on the transmission of mmW signals beyond the mid-boundary of the EHF frequency band of the 5G frequency spectrum (Bessemoulin et al., 2005; Zibar et al., 2011), the link length reduces along with the augmentation of power penalty to achieve the threshold BER, as shown in Figure 8.7 and Figure 8.8. An SNR penalty of ≈25 dB and ≈34 dB is required to attain the BER of $10^{-3}$ at 55.6 GHz over a wireless link of 10 m and 50 m, respectively. For the successful transmission at 110 GHz, a power penalty of ≈35 dB and ≈45 dB is required over the demonstrated link lengths due to high fading in this frequency band. Thus, this work shows the feasibility of using the proposed laser for realizing the RoF transmission systems over a wide span of mmW band with a possibility of attaining an effective data rate of ≈100 Gbps over a wireless link up to ≈10 m using the existing state-of-the-art 75–110 GHz antenna technology capable of providing a combined antenna gain of ≥48 dBi. Moreover, the spatial diversity and beamforming techniques may play a significant role in achieving high transmission data rates in non-LOS environments at minimal power requirements.

### 8.3.2 TRANSMISSION OF MMW WAVES OVER A FREE-SPACE OPTICS (FSO) LINK

We have also designed the optical communication system to transmit a data rate of 10 Gb/s (16-QAM signals) over FSO using four different carrier frequencies (56.7, 77.6, 86.2, and 110 GHz) that are generated by the DWFL. The schematic diagram for carrying out the transmission of 16-QAM signals over the FSO link is shown in Figure 8.9. This high data rate has been transmitted over the FSO link using an Optisystem photonic module to validate the potential of the proposed laser. The dual-wavelength optical signals, for instance, 0.46 nm (~56.7 GHz) are separated using DWDM (1:2 splitter) into two wavelengths (1558.94 nm and 1559.40 nm). In the first arm of the splitter, the 16-QAM are optically modulated over the optical frequency carrier of 1559.40 nm using a MZM. A suitable optical BPF has been used with a center wavelength of 1559.40 nm to filter out the high sideband and only obtain the lower sideband of the QAM signals i.e., LSB-QAM. This LSB-QAM signal is applied to another DWDM (2:1 combiner) to combine with the unmodulated output signal in

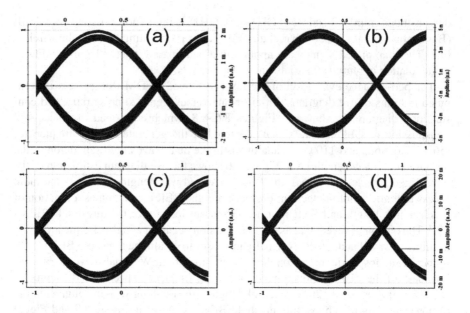

**FIGURE 8.6** Eye diagram estimation over a fiber link of 20 Km at wavelength spacing of (a) 0.46 nm (b) 0.63 nm (c) 0.70 nm, and (d) 0.89 nm. (Adapted from Kbashi et al., 2021b.)

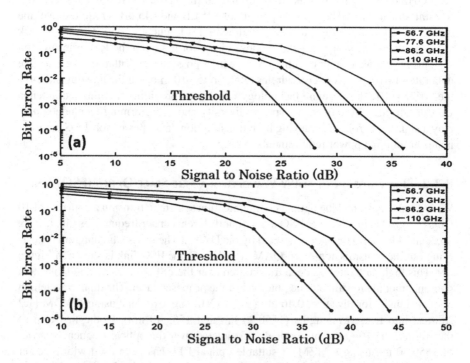

**FIGURE 8.7** BER measurements versus SNR penalty at varied mmW signals over the wireless link of (a) 10 m, and (b) 50 m. (Adapted from Kbashi et al., 2021b.)

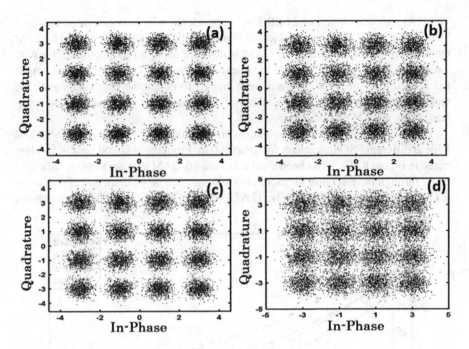

**FIGURE 8.8** Constellation diagrams over wireless link of 10 m at (a) 56.7 GHz @ SNR = 25 dB. (b) 77.6 GHz @ SNR = 29 dB. (c) 86.2 GHz @ SNR = 31.5 dB, and (b) 110 GHz @ SNR = 35 dB. (Adapted from Kbashi et al., 2021b.)

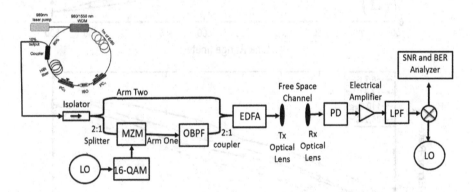

**FIGURE 8.9** Schematic diagram of the 5G transmission over FSO link using erbium-doped DWFL.

the second arm of DWDM. The obtained LSB-QAM modulated signals at the output of the combiner are transmitted over 20 Km of SMF and then to the FSO wireless channel of 500 m. The LSB-QAM signals are amplified before transmitting through the FSO link by using low-noise EDFA (gain = 17 dB, NF = 5 dB) to restore the optical power budget. The proposed transmitter and receiver telescopic lens have aperture diameters of 5 cm and 15 cm, respectively, with a beam divergence of 2 mrad.

The FSO link attenuation of 21 dB/Km is considered with fiber-telescope coupling losses equal to 1 dB; geometrical losses equal to 2 dB; mispointing equal to 5 dB. At the receiver side, the received 16-QAM is detected using a PIN photodetector with a responsivity of 0.8 A/W. Due to the attenuating wireless link, an electrical amplifier is used to amplify the received weak signal. To obtain the baseband signal, the received electrical signals are passed out of a low pass filter (LPF) and are demodulated using a QAM demodulator.

The system transmission performances are considered by measuring the SNR and BER at different FSO link length for up to 500 m. From the obtained results, as illustrated in Figure 8.10a, it has been shown that the 16-QAM data are transmitted effectively through the FSO link for more than 500 m, ≈350 m, ≈275 m, and 200 m using the carrier frequencies of 56.7 GHz, 77.6 GHz, 86.2 GHz, and 105 GHz, respectively.

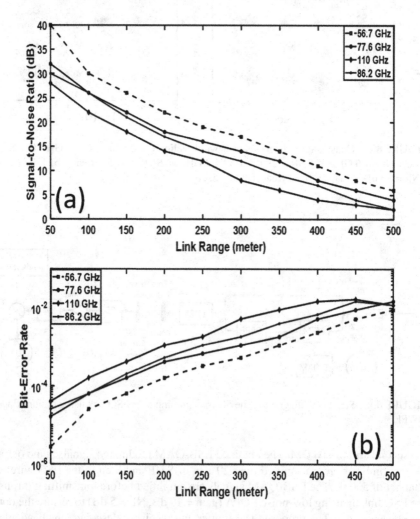

**FIGURE 8.10**    (a) SNR and (b) BER measurements of the 5G transmission driven by DWFL.

Additionally, BER consideration as a function of the FSO link is also carried out as shown in Figure 8.10b, which reveals that the achievable transmission link reduces to obtain the acceptable BER of $10^{-3}$ target at a higher mmWaves frequencies. These results are computed based on the SNR threshold value of 15 dB for all four carrier frequencies showing clearly that the 16-QAM data signals of 10 Gb/s are transmitted successfully.

## 8.4 DWFL FOR LIDAR APPLICATIONS

### 8.4.1 EXPERIMENTAL RESULTS AND DISCUSSION

The designed and demonstrated DWFL (Kbashi et. al. 2021) have been utilized to develop a linear frequency-modulated continuous-wave (FMCW) photonics-based radar system to compute the range and speed of several vehicles unambiguously in a modeled traffic scenario via the co-simulation of *MATLAB*™ and *OptiSystem*™ software. After a precise adjustment of the polarization controllers in the DWFL setup, a dual-wavelength at $\lambda_1 = 1559.50$ nm and $\lambda_2 = 1560.125$ nm is attained and generated corresponding mmW waves of 77 GHz as shown in the RF spectra in Figure 8.11.

Figure 8.12 depicts the basic idea of a DWFL-driven linear FMCW photonics-based radar system that employs the proposed laser to utilize as an RF carrier and a light-based carrier-source simultaneously. A linear frequency-modulated signal (LFM) is generated using the proposed laser output (at Port 1) tuned to 77 GHz and a sawtooth pulse generator. The generated LFM radar signals are further optical modulated over the laser output (at Port 2) via an external dual-electrode MZM after passing through a tunable bandpass filter centered at 1559.5 nm. The LFM signal is fed directly to the first arm of the modulator. It is applied to the second arm of the modulator through a phase-shifter. The unwanted higher-order sidebands are suppressed by adjusting the bias voltages of the external modulator (Sharma and Sergeyev, 2020; Sharma et al., 2021).

This optical treatment extends the spectral width of the LFM radar signals which helps in improving the radar resolutions. These spectrally broadened signals are transmitted toward the target objects using an antenna module after photo-detection through a narrowband line-of-sight (LOS) channel. The channel is modeled as per ITU-R P.838-3 and ITU-R P.840-6 recommendations (ITU-R P.838-3, 2005; ITU-R P.840-6, 2013), including the water vapor-density, dry-air pressure, liquid water-density, and rainfall rate, to compute the impact of weather perceptions as these degrade the radar signals significantly at mmW frequencies. The channel is developed using the phased-array tool of *MATLAB*™ and is further co-simulated with the well-known photonics-based simulation tool, i.e., *OptiSystem*™ to investigate the DWFL-driven photonics-based radar.

A traffic scenario is also developed, including large-and small-sized automotive mobile targets, i.e., cars, trucks, motorbikes, bicycles, and pedestrians traveling at different relative velocities (1.11 m/s, 5.55 m/s, 8.33 m/s, 20.83 m/s, and 26.38 m/s respectively) at diverse target ranges (490 m, 400 m, 350 m, 300 m, and 250 m respectively) upfront of the observatory vehicle moving at 100 km/hr assuming all the target vehicles moving toward it. All the target objects are defined by their

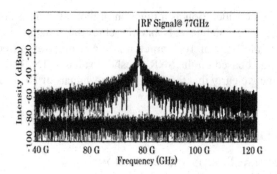

**FIGURE 8.11**　RF signal at 77 GHz after proper tuning of the laser at wavelength spacing of 0.625 nm.

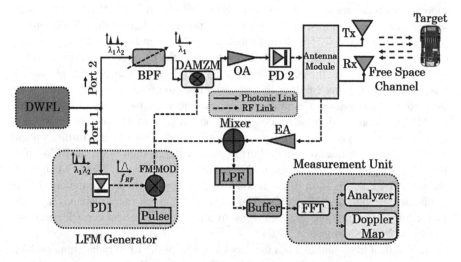

**FIGURE 8.12**　Schematic diagram of DWFL-driven LFMCW-PHRAD system.

associated radar cross-section (*RCS*) at 77 GHz as per Sharma et al. (2021) and Yamada (2001). A down-conversion of the reflected echoes with the instantaneous transmitted signal is carried out to measure the beat frequency. A down-conversion of the reflected echoes with the instantaneous transmitted signal is carried out to measure the beat frequency and, thus, the detection range. The received dechirp signals for 128 sweeps are stored in a buffer to retrieve the Doppler shift. Then, a two-dimensional fast Fourier transformation with the Hanning window is applied to measure the range and speed concurrently of the illuminated targets corresponding to the retrieved beat frequency and Doppler-shift information at a minimal false alarm rate (Sharma 2020; Sharma et al., 2021).

Under a clear-weather scenario shown in Figure 8.13, the LOS channel is modeled with dry air pressure = 101, water-vapour density = 0, liquid water-density = 0, and rainfall rate = 0 mm/hr. The beat-frequency is measured as 267.19 MHz, 245.51 MHz, 218.55 MHz, 136.52 MHz, and 191.02 MHz with signal-intensity of −103.83 dBm/Hz,

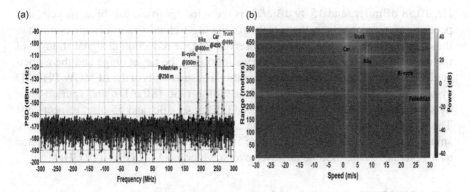

**FIGURE 8.13** Measurements in terms of (a) PSD and (b) Doppler mapping under clear-weather scenario for DWFL-driven LFMCW-PHRAD system.

−111.39 dBm/Hz, −112.512 dBm/Hz, −122.5 dBm/Hz, and −112.06 dBm/Hz for truck, car, bike, bicycle, and pedestrian, respectively. For severe weather situations, including the influence of heavy rain and heavy fog, the LOS channel is modeled with a liquid water density = 7.5, liquid water density = 2.5, and rainfall rate = 50 mm/hr (Yamada, 2001; Sharma, 2020; Sharma et al., 2021). The impact of smoke, atmospheric gasses, and dust particles are not considered for all the considered scenarios in this work. For the severe weather scenario shown in Figure 8.14, the beat-frequency is measured with the signal intensity of −124.95 dBm/Hz, −1130.66 dBm/Hz, −126.68 dBm/Hz, −133.38 dBm/Hz, and −127.15 dBm/Hz for the targets under observations in the same order of the clear-weather scenario. The measured beat frequency of the demonstrated DWFL-driven LFMCW-PHRAD matches the theoretical measurements with a marginal frequency error of a few kHz. The attained outcomes reveal that heavy fog and rain have a substantial influence on the detection range and the echo strength. A difference in signal fading of 21.12 dBm/Hz, 19.27 dBm/Hz, 14.17 dBm/

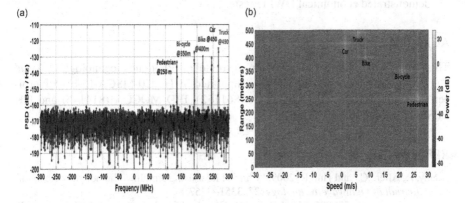

**FIGURE 8.14** Measurements in terms of (a) PSD and (b) Doppler mapping under the collective influence of environmental factors.

Hz, 10.88 dBm/Hz, and 15.09 dBm/Hz is recorded for truck, car, bike, bicycle, and pedestrians, respectively, in the presence of the fog + cloud + rain scenario contrast to the clear-weather scenario. It is also observed that the reduction in signal strength of the received reflected radar signals varies following the target range and the associated RCS of the illuminated targets. The other measurements of the LFMCW-PHRAD system, like SNR, signal-to-noise and distortion ratio (SINAD), and spurious-free dynamic range (SFDR), are recorded as 43 dBm, 43 dBc, and 55 dBc, respectively. The recorded observations validate the ability of the demonstrated laser to realize smart transportation systems and diverse 5G-supported microwave-photonics systems as it generates stable millimeter waves over a wider band (12 GHz–110 GHz).

## 8.5 CHAPTER CONCLUSIONS

In this chapter, we demonstrate the tunable dual-wavelength erbium-doped fiber laser to generate millimeter-wave signals over a wide range of 10 GHz to 110 GHz using an erbium-doped and high birefringence fibers in NPR ring configuration. This DWFL source has been successfully realized for transmission of 16-QAM with the data rate of 10 Gb/s over RoF and FSO links with an acceptable SNR and BER for all four different generated mm-waves carrier frequencies (56.7 GHz, 77.6 GHz, 86.2 GHz, 110 GHz).

We have also utilized the DWFL laser for estimating the target range and velocity of multiple RCS-defined automotive targets by developing an *LFMCW-PHRAD* in the presence of environmental fluctuations via *MATLAB*™ and *OptiSystem*™ software. Due to attaining a uniform wavelength spacing, along with low phase noise, the established DWFL-driven *LFMCW-PHRAD* system can provide phase-stable and spectrally-broadened optical-modulated radar signals for achieving high-imaging resolutions with high accuracy. As the demonstrated photonics-based radar system is established in a simulative environment using the experimentally designed DWFL laser, an end-to-end fully experimental system may be implemented in the future to realize futuristic 5G-supported smart transportation systems/networks, tunable multiband photonics-based radar systems and automotive photonics radar services using the demonstrated economical DWFL system.

## ACKNOWLEDGMENTS

This work was carried out at Aston Institute of Photonic Technologies, Aston University, Birmingham, UK, and was supported by H2020-MSCA-IF-EF-ST project (840267), H2020 ETN MEFISTA (861152), EID MOCCA (814147) and UK EPSRC project EP/W002868/1.

## REFERENCES

Adany, P., Allen, C., and Hui, R. 2009. "Chirped lidar using simplified homodyne detection." *Journal of Lightwave Technology*, 27, 3351–3357.

Bessemoulin, A. et al. 2005. "High gain 110-GHz low noise amplifier MMICs using 120 nm metamorphic HEMTs and coplanar waveguides." *European Gallium Arsenide and Other Semiconductor Application Symposium, GAAS*, Paris, France, pp. 77–80.

Dennis, M. L., Putnam, M. A., Kang, J. U., Tsai, T.-E., Duling, I. N., and Friebele, E. J. 1997. "Grating sensor array demodulation by use of a passively mode-locked fiber laser." *Optics Letters*, 22, 1362–1364.

Fortier, T. M., Kirchner, M. S., Quinlan, F., Taylor, J., Bergquist, J., Rosenband, T., et al. 2011. "Generation of ultrastable microwaves via optical frequency division." *Nature Photonics*, 5, 425–429.

Gao, O'Sullivan M., and Hui, R. 2012. "Complex-optical-field Lidar system for range and vector velocity measurement." *Optics Express*, 20, 25867–25875.

Hu, G., Mizuguchi, T., Oe, R., Nitta, K., Zhao, X., Minamikawa, T., Li, T., Zheng, Zh., and Yasui, T. 2018. "Dual terahertz comb spectroscopy with a single free-running fiber laser." *Scientific Reports*, 8, 11155.

International Telecommunication Union-R P.838-3, "Specific attenuation model for rain for use in prediction methods," (2005).

International Telecommunication Union-R P.840-6, "Attenuation due to clouds and fog" (2013).

Kbashi, H. J., Sergeyev, S. V., Mou, Ch., Garcia, A. M., Al Araimi, M., Rozhin, A., Kolpakov, S., and Kalashnikov, V. 2018. "Bright-dark rogue waves." *Annalen der Physik*, 530, 1700362.

Kbashi, H. J., Sharma, V., and Sergeyev, S. 2021a. "Phase-stable millimeter-wave generation using switchable dual-wavelength fiber laser." *Optical Lasers Engineering*, 137, 106390.

Kbashi, H. J., Sharma, V., and Sergeyev, S. 2021b. "Dual-wavelength fiber-laser-based transmission of millimeter waves for 5G-supported radio-over-Fiber (RoF) links." *Optical Fiber Technology*, 65, 2021, 102588.

Kbashi, H. J., Sharma, V., and Sergeyev, S. 2021c. "Transmission of 5G using tunable dual-wavelength fiber laser." *2021 Conference on Lasers and Electro-Optics Europe (CLEO), Microwave Photonics (ci_3)*, Munich Germany, 21–25 June 2021, ISBN: 978-1-6654-1876-8.

Khattak, G. Tatel, and Wei, L. 2018. "Tunable and switchable EDF laser using a multimode-fiber based filter." *Applied Sciences*, 8, 1135.

Kolpakov, S. A., Kbashi, H. J., and Sergeyev, S. V. 2016. "Slow optical rogue waves in a unidirectional fiber laser." *2016 Conference on Lasers and Electro-Optics (CLEO)*, San Jose, California United States, 5–10 June 2016, ISBN: 978-1-943580-11-8. Science and Innovations, pp. JW2A-56.

Lian, Y. et al. 2017. "Switchable multi-wavelength fiber laser using erbium-doped twin-core fiber and nonlinear polarization rotation." *Laser Physics Letters*, 14, 055101.

Morris, O. J., Wilcox, K. G., Head, C. R., Turnbull, A. P., Mosley, P. J., Quarterman, A. H., Kbashi, H. J., Farrer, I., Beere, H. E., Ritchie, D. A., and Troppe, A. C. 2012. "A wavelength tunable 2-ps pulse VECSEL," *2012 Proceedings SPIE 8242, Vertical External Cavity Surface Emitting Lasers (VECSELs) II*, 824212, 14 February 2012, San Francisco, California, United States, https://doi.org/10.1117/12.908337.

Ramasubramanian, K., and Ramaiah, K. 2018. "Moving from legacy 24 GHz to state-of-the-Art 77-GHz Radar."*ATZ Elektronik Worldwide*, 13, 46–49.

Sergeyev, S. V., Kbashi, H., Tarasov, N., Loiko, Yu and Kolpakov, S. A. 2017. "Vector-resonance-multimode instability." *Physicsal Review Letters*, 118, 033904.

Sharma, V., Kbashi, H. J., and Sergeyev, S. 2021. "MIMO-employed coherent photonic-radar (MIMO-Co-PHRAD) for detection and ranging." *Wireless Networks*, 27, 2549–2558.

Sharma, V. and Kumar, L. 2020. "Photonic-radar based Multiple-Target Tracking under complex traffic-environments", *IEEE Access*, 8, 225845–225856.

Sharma, V., and Sergeyev, S. 2020. "Range detection assessment of photonic-radar under adverse weather perceptions.", *Optics Communications*, 472, 125891.

Sharma, V., Sergeyev, S., Kumar, L., and Kbashi, Hani. 2020. "Range-speed mapping and target-classification measurements of automotive targets using photonic radar", *Optical and Quantum Electronics*, 52 (438), 3–18.

Sibbett, W., Lagatsky, A. A. and Brown, C. T. A. 2012. "The development and application of femtosecond laser systems," *Optics Express, 20,* 6989–7001.

Tsai, Ch. T., Li, Ch., Lin, Ch., Ch, T. L. and Lin, G. 2018. "Long-reach 60-GHz MMWoF link with free-running laser diodes beating," *Scientific Reports, 8,* 13711.

Udem, T., Holzwarth, R. and Hansch, T. W. 2002. "Optical frequency metrology." *Nature, 416,* 233–237.

Yadav, A., Kbash, H. J., Kolpakov, S., Gordon, N., Zhou, K., and Rafailov, E. U. 2017. "Stealth dicing of sapphire wafers with near infra-red femtosecond pulses." *Applied Physics A, 123,* 369.

Yamada, N. 2001. "Three-dimensional high-resolution measurement of radar cross section for car in 76 GHz band." *R&D Review of Toyota Central R&D Labs, 36* (2), 1–2.

Yan, N. et al. 2017. "Tunable dual-wavelength fibre-laser with unique-gain system-based on in-fibre acousto-optic mach–zehnder interferometer." *Optics Express, 25*(22), 27609.

Zhang, F. et al. 2017. "Photonics-based real-time ultra-high-range-resolution radar with broadband signal generation and processing," *Scientific Reports, 7* (1), 1–8.

Zhou, J. et al. 2015. "Dual-wavelength single-longitudinal-mode fibre laser with switchable wavelength spacing based on a graphene saturable absorber." *Photonics Research, 3*(2), A21–A24.

Zhu, D.-Q. and Li, P.-B. 2017. "Preparation of entangled states of microwave photons in a hybrid system via the electro-optic effect." *Optics Express, 25* (23), 28305–28318.

Zibar, D., Caballero, A., Yu, X., Pang, X., Dogadaev, A. K. and Monroy, I. T. 2011. "Hybrid optical fibre-wireless links at the 75–110 GHz band supporting 100 Gbps transmission capacities." *2011 International Topical Meeting on Microwave Photonics jointly held with the 2011 Asia-Pacific Microwave Photonics Conference,* Singapore, 2011, pp. 445–449, doi:10.1109/MWP.2011.6088767.

# Index

## A

abnormal dispersion, 189–190
additive white Gaussian noise (AWGN), 211
all-fiber Lyot filter (AFLF), 83
Amplified Spontaneous Emission (ASE), 107
autocorrelation (AC), 76
auto-correlator (Pulsecheck), 2
Autonomous Vehicles (AVs), 206

## B

Bar-Eli effect, 39
Benjamin-Feir instability (BFI), 23
bi-directional propagation, 125–127
bound state soliton, 11, 12, 21, 22, 24–27
bright-dark rogue waves (BDRWs), 21

## C

color domain (CD), 110
carbon nanotubes (CNT), 2, 3, 9, 18, 31, 65, 80, 99, 134, 139, 140, 187, 199
cavity-space multiplexing, 125–127
clockwise (CW), 189
coherent cross-polarization coupling (XPC), 145
    vector solitons formed under, 145–153
        coherently coupled vector black solitons, 147–148
        coherently coupled vector black-white solitons, 148–149
        coherently coupled vector bright solitons, 145–146
color domain (CD), 110
color domain walls, (CDW), 103, 109–119, 121
comb generation, dual frequency, 124
Complex Ginzburg-Landau Equations (CGLEs), 107
continuous waves (CW), 155
conventional soliton (CS), 80
counterclockwise (CCW), 189
cross-phase modulation (XPM), 144

## D

degree of polarization (DOP), 4, 18, 91, 208, 210
dense wavelength division multiplexing (DWDM), 81
dispersion compensating fiber (DCF), 154, 156
dissipative parametric instability (DPI), 23
dissipative solitons (DSs), 80
dissipative vector solitons (DVSs), 1

dispersion management
    abnormal dispersion, 189–190
    near-zero net dispersion, 197–199
    normal dispersion, 190–197
    partially mode-locked fiber lasers, 199–203
dispersive Fourier transform (DFT), 133
dissipative parametric instability (DPI), 23
dissipative soliton resonance (DSR), 170
    typical pulse spectrum operating, 178–179
    vector nature, 179–181
dissipative solitons (DSs), 80
dissipative vector soliton (DVS), 194
dissipative vector solitons (DVSs), 1
domain walls, 103, 104, 109, 110, 115, 116, 118–121, 166
    color, 109–119
    polarization, 103, 104, 107, 109, 110, 116, 119, 120, 166
double scroll polarization attractor (DSPA), 57
driver assistance system (DAS), 206
dual color domain (DCD), 113
dual-output vector soliton fiber lasers, 123–124
    comb generation, 124
    polarization-multiplexed dual output pulse trains, 128–133
    pulse trains, 124–128
dual-wavelength fiber-laser (DWFL), 205–207
    characterization, 207–210
    experimental setup, 207–210
    LiDAR applications, 217–223
    transmission applications, 211–217
dual-wavelength operation, 127–128

## E

Eickhoff, W., 97
Er-doped fiber, 2, 18, 21, 23, 26, 27, 36, 49, 57, 75–78, 80, 84, 86, 88, 89, 91–93, 96, 98, 99, 126, 128, 130, 134, 207, 208
Er-doped fiber laser, 2, 3, 26, 49, 57, 75–78, 80, 83, 86, 88, 89, 91–93, 96, 98, 99, 126, 127, 130, 134
    vector model, 2, 23, 30, 39, 49, 53, 55
        semiclassical equations, 1, 30
experiments
    vector soliton rain, 1, 18, 19, 26, 44, 64, 67
    vector resonance multimode instability, 1, 23, 26, 49, 53, 54, 67, 199, 210
    vector harmonic mode-locking, 1, 26, 53, 67
    fundamental soliton polarization dynamics, 1, 2
    vector multipulsing soliton dynamics, 1, 5

**F**

Fabry-Perot interferometer (FPI), 79
femtosecond laser inscription, 69, 74, 75, 93
figure-eight fiber lasers
    multi-soliton pattern vector nature, 171–177
        cluster/soliton flow, 175–177
        fundamental vector soliton, 172
        random static distribution, 172–175
    noise-like pulse trapping, 181–186
    vector dissipative soliton resonance, 178–181
    vector solitons in, 169–188
    working principle, 170–171
four-wave mixing (FWM), 130, 144, 149, 199
Fourier transform, 10, 64, 65, 133, 136, 140, 198, 218
free spectral range (FSR), 83
free-space optics (FSO), 206, 213–217
frequency-modulated continuous-wave (FMCW), 217
full-width half-maximum (FWHM), 2
fundamental soliton polarization dynamics
    experimental set-up, 1, 2
    experimentally observed attractors, 1, 4

**G**

Gao, Lei, 191
GHz harmonic mode-locked fiber laser, 69, 88–89
Ginzburg-Landau equation (GLE), 191
group velocity locked bright solitons, 159–160
group-velocity-locked solitons (GVLS), 123
group-velocity-locked vector solitons (GVLVSs), 170
group velocity dispersion (GVD), 2, 21, 104, 151, 154, 178, 191

**H**

harmonic mode-locking, 1, 18, 26, 65, 67, 85, 109, 114, 121, 190, 192
high birefringence (HiBi), 207
High Nonlinear Fiber, 95, 110

**I**

in-fiber polarization beam splitter, 69, 89, 91
in-line polarimeter (Thorlabs, IPM5300), 2
incoherent cross-polarization coupling (XPC)
    vector solitons formed under, 149–153
        incoherently …, 150–151
        incoherently …, 149–150
        incoherently …, 151–153
incoherently coupled dark-bright solitons, 161–164
incoherently coupled gray solitons, 160
independent optical isolator (ISO), 207
interplay between birefringence and polarization
    hole burning, 41–43

**K**

Kbashi, H., 199
Kurkov, A. S., 98

**L**

laser diode, 2, 23, 24, 27, 28, 171, 207, 223
LiDAR, DWFL, 217–223
line-of-sight (LOS), 217
linearly birefringence, SMFs, 143–144
    experimental studies, 153–164
        coherently coupled vector solitons, 155–159
        experimental setup, 153–155
        incoherently coupled…, 159–164
    outlook, 164–167
    pulse propagation, 144–145
    XPC vector solitons, 145–153
long period grating (LPG), 71, 97, 98
long wavelength (LW), 113
low pass filter (LPF), 216
low-noise amplifier (LNA), 211

**M**

Mach Zehnder modulator (MZM), 211
Mach-Zehnder interferometer (MZI), 79
mode-locked fiber lasers
    based on 45° TFG, 69, 72, 75, 96
        GHz harmonic, 69, 88
        in-fiber polarization beam splitter, 69, 89, 91
        multi-wavelength, 69, 79, 81–83, 99, 127, 135, 140, 187, 222
        pulse state switchable, 69, 85, 87, 88, 99
        stretched-pulse, 69, 77–79, 85, 133, 140
        wavelength tunable/switchable, 69, 79, 82
    developing polarizing fiber grating based, 69–102
    dispersion management, 189–203
    dual-output vector soliton fiber lasers, 123
    polarizing fiber grating based, 5, 69, 72
    polarization dynamics of, 1–68
    polarization-color domains, 103–121
modulation instability (MI), 23–26
multi-wavelength mode-locked fiber laser, 69, 79, 81

**N**

near-zero net dispersion, 197–199
noise-like pulse trapping, 181–186
    from conventional soliton to noise-like pulse, 181–183
    wavelength shift, 184–186
noise-like pulses (NLP), 85

nonlinear amplified loop mirror, 18, 70, 170
nonlinear optical loop mirror, 70, 97, 170
nonlinear polarization evolution, 70, 134
nonlinear polarization rotation, 9, 18, 23, 27, 49,
        98–100, 123, 125, 17, 129, 134, 170,
        189, 202, 207, 222
nonlinear Schrödinger equation, 21, 63, 103, 166,
        169, 186
normal dispersion, 197

O

OFS TruePhase® IPLM, 5, 9, 12
optical bandpass filter, 211
optical isolator, 2, 3, 24, 27, 28, 171, 207
optical spectrum analyzer, 2, 3, 208
optical tunable filter, 112, 113
OptiSystem, 211, 217
Ortega, B. L., 71, 100
oscilloscope, 2
output coupler (OUTPUT C), 2, 21, 27
output lasing, 2, 38

P

partially mode-locked (PML), 199–203
period doubling, 129–133
photonic crystal fiber (PCF), 71
Poincaré sphere, 4, 6–9, 11–18, 20, 22, 24, 38, 42,
        43, 45, 46, 48, 55–58, 60, 193, 194,
        197–201, 210
polarization beam splitter (PBS), 189
polarization controller, 2–5, 9, 11, 12, 18, 24–29,
        36, 37, 48, 54, 55, 64, 76, 105, 109,
        110, 112, 113, 115, 128–132, 154, 155,
        170, 171, 189, 197
polarization coupling coefficient (PCC), 132
polarization domain wall, 68, 103, 104, 106, 107,
        109, 110, 116, 119–121, 166
polarization dynamics, mode-locked fiber lasers,
        1–2
    bound state stations, 9–17
    EDFL vector model, 30–37
    fundamental soliton polarization dynamics,
        2–5
    interplay, 41–43
    self-pulsing, 57–62
    spiral polarization attractor, 37–41
    vector bright-dark rogue waves, 21–23, 47–49
    vector harmonic mode-locking, 26–29, 53–56
    vector multipulsing soliton dynamics, 5–9
    vector resonance multimode instability, 23–26,
        49–53
    vector soliton rain, 18–20, 44–47
polarization evolution frequency (PEF), 129
polarization extinction ratio, 70, 71, 73
polarization hole burning, 1, 41, 43–45, 190
polarization maintaining (PM), 127

Polarization Mode Dispersion, 71, 107, 126,
        128, 144
polarization rotation vector soliton cluster
        (PRVSC), 175–177
polarization rotation vector solitons (PRVS), 123
polarization rotation vector solitons, fundamental
        mechanism of, 129
polarization-dependent loss, 71, 96, 105, 107–110,
        118
polarization-independent isolator (ISO), 171
polarization-locked vector soliton (PLVS), 169
polarization-locked vector solitons (PLVS), 123
polarization-maintaining, 5, 71, 97, 98, 110, 136,
        137, 139, 197
polarization-maintaining fiber (PMF), 110
pulse state switchable mode-locked fiber laser,
        85–88
pulse train, dual output
    dynamics of, 133–135
    polarization-multiplexed, 128–133
pulse trains, dual output, 124–125
    bi-directional propagation, 125–127
    cavity-space multiplexing, 125–127
    dual-wavelength operation, 127–128
    spectral properties, 127–128
pump laser diode, 27
pump-to-signal intensity noise transfer (PSINT), 57

R

radar cross-section (RCS), 218
radio access unit (RAU), 211
radio-over-fiber (RoF), 206, 211–213
radiofrequency (RF), 25, 124
Risken-Nummedal-Graham-Haken, 23, 65, 68
rogue wave, 1, 21, 22, 47, 48, 57, 63–65, 67, 68,
        134, 137, 193, 199, 203, 222

S

saturable absorber (SA), 2, 70, 109, 191
self-induced polarization change (SPM), 103
self-phase modulation (SPM), 144
self-pulsing, 1, 57, 61–64, 67, 120
self-Q-switching, 57, 64
semiconductor saturable absorber mirror
        (SESAM), 170, 194
Sergey, S. V., 197
Shilnikov theorem, 40, 58
short wavelength (SW), 113
short-range coordination order (SRCO), 57
signal-to-noise and distortion ratio (SINAD), 221
signal-to-noise ratio (SNR), 27, 76
single color domain (SCD), 110
single-mode fiber (SMF), 70, 156, 169
    theoretical studies, 144–153
    vector solitons formed in, 143–167
single-wall carbon nanotubes, 18, 31, 139, 140

slandered telecommunication fiber (SMF), 207
spectral properties, 127–128
spiral polarization attractor, 1, 37–41
spurious-free dynamic range (SFDR), 221
state of polarization, 1, 6, 7, 18, 190, 207, 208
stretched-pulse mode-locked fiber laser, 69, 78

T

theories
    vector soliton rain, 1, 18, 19, 26, 44, 64, 67
    vector bright-dark rogue waves, 1, 21, 47
    vector resonance multimode instability, 1, 23,
        26, 49, 53, 54, 67, 199, 210
    vector harmonic mode-locking, 1, 26, 53, 67
Third Order Dispersion (TOD), 110
tilted fiber gratings, 5, 71, 99, 101
    Fabrication and Characterization, 69, 72
    femtosecond laser inscription, 69, 74, 75, 93
    mode-locked fiber lasers based on 45° TFG,
        69, 72, 75, 96
    GHz harmonic, 69, 88
    in-fiber polarization beam splitter, 69, 89, 91
    influence, 69
    multi-wavelength, 69, 79, 81–83, 99
    pulse state switchable, 69, 85, 87, 88, 99
    stretched-pulse, 69, 77–79, 85
    wavelength tunable/switchable, 69, 79, 82
time-bandwidth product (TBP), 76
transmission applications, DWFL
    free-space optics, 213–217
    radio-over-fiber, 211–213
triple color domain, 113, 114

V

vector black solitons, 156–157
vector black-white solitons, 157–159
vector bright solitons, 155
vector bright-dark rogue waves, 1, 21, 47
vector harmonic mode-locking, 1, 23, 53, 67
vector multipulsing soliton dynamics, 1, 5
vector resonance multimode instability, 1, 23, 26,
        49, 53, 54, 67, 199, 210
vector soliton rain, 1, 18, 19, 26, 44, 64, 67
vector solitons
    figure-eight fiber lasers, 169–188
    formed in linearly birefringent SMFs,
        143–167
vector-resonance-multimode instability (VRMI),
        199

W

wavelength division multiplexing (WDM), 2, 3,
        81, 207
wavelength tunable/switchable mode-locked fiber
        laser, 79–85

X

X-ray-absorption fine structure spectroscopy
    (XAFS), 57

Y

Yan, Z. Y., 68